# Viruses vs. Superbugs
## A solution to the antibiotics crisis?

Thomas Häusler

Translated by Karen Leube

First published 2006 by
Macmillan
Houndmills, Basingstoke, Hampshire RG21 6XS and
175 Fifth Avenue, New York, N.Y. 10010
Companies and representatives throughout the world

ISBN-13: 978–1–4039–8764–8
ISBN-10: 1–4039–8764–5

This book is printed on paper suitable for recycling and
made from fully managed and sustained forest sources.

A catalogue record for this book is available from the British Library.

A catalog record for this book is available from the Library of Congress.

10  9  8  7  6
15  14  13  12  11  10  09  08  07  06

Printed and bound in China

*For Susanne and Julia*

# contents

# list of figures

# foreword

Just imagine life without antibiotics. It would be like it was 100 years ago, when pneumonia and tuberculosis were the most frequent causes of death, and the risk of infection turned a simple appendectomy into a dangerous operation.

Luckily we do have antibiotics. However, they are becoming increasingly ineffective. Doctors are more and more frequently confronted with infections they can't do anything about because the bacteria have become resistant. This has dire consequences for patients. Many end up living with a chronic infection for years on end, some are forced to become amputees and yet others succumb to the infections.

The crisis affects people in both industrialized and developing countries. In the US and the UK, the bug *Staphylococcus aureus* is wreaking havoc. Forty to fifty per cent of infections that people contract in hospitals are resistant to more than one antibiotic. The developing countries are groaning under the burden of tuberculosis, which claims the lives of 2 million victims throughout the world every year. The increase in multi-resistant TB is especially alarming. Treating it costs 100 times more than treating the regular form, making a cure unafford-able for many people in impoverished countries. And these are only two examples.

Despite this, many pharmaceutical companies have stopped developing antibiotics. They see the financial risk as too big and potential profits too skimpy. This has led to very few new drugs for fighting bacterial infections being launched in recent years. A survey of 11 large pharmaceutical companies revealed that of 400 substances they were developing, only 5 were anti-bacterial drugs.

What can be done about the resistance crisis? One thing is needed for sure: new drugs. One of them could be bacterio-phages, viruses that attack bacteria without harming people. So-called bacteriophage therapy had its heyday from 1920 to 1940, before it was pushed aside by penicillin. The former Soviet Union is the only place it continues to be used today. Most Western doctors do not even know that this method exists.

However, there are some scientists who have resumed research on bacteriophage therapy, and that's a good thing. We need to pursue any and every approach that can contribute to solving the resistance crisis. Bacteriophage therapy may prove to be a particularly worthwhile area of research. Its long history provides a large stock of knowledge that is freely accessible. Determined researchers now need to use this as a starting point and work out how to turn bacterio-phages into drugs that meet today's standards. This would be best carried out in cooperation with science departments at universities, along with private companies and non-profit foundations that support the projects. This is exactly the goal of the Foundation for Fatal Rare Diseases. The foundation supports the development of drugs for neglected infectious and pulmonary diseases and is especially committed to helping affected patients who have not been in the public eye, particularly those in Africa and India.

Thomas Häusler's remarkable book plays a central part in this scheme, because it acquaints the public, researchers and deci-sion makers with a therapy that has the potential to someday heal many patients who cannot be helped at the moment. This is why the Foundation for Fatal Rare Diseases is supporting the realization of this English edition. The fact that the author writes about bacteriophage therapy in the form of such a gripping story makes reading it all the more exciting.

Vaduz, October 2005
Vera Cavalli, Dorian Bevec and Fabio Cavalli
Founders of the Foundation for Fatal Rare Diseases

# preface

Why should anyone be interested in an old cure that hasn't been used in the West for 50 years? It's a method that many doctors aren't even aware of today. The most telling answer to this question came when I received a call in my office from a man one Friday morning in January 2001. It was the day after my article on the Eliava Institute in Georgia had appeared in the German weekly newspaper *Die Zeit*. In the article I had described how this old remedy – phage therapy – had survived in the impoverished country.

Phages are viruses that attack and kill bacteria but not people. Since Stalin's days, doctors in Russia and Georgia have been using phages to cure bacterial infections. In the West this method was also once popular but, in contrast to the Soviet Union, the triumph of penicillin pushed phage therapy aside here after 1940. The Eliava Institute in Tbilisi, Georgia is a place where phage therapy survived even after the collapse of the Soviet Union. It looks back on a glorious 80-year history. However, because of Georgia's economically and politically precarious situation, it is experiencing a gloomy present. From the point of view of today's science, it is unclear how effective phages are in fighting infection. This is because the studies carried out by early pioneers and Soviet researchers do not meet today's standards. All this was in my article in *Die Zeit*.

The caller explained that he had read the article. He was calling directly from the hospital and appeared to be under a great deal of pressure. Not mincing words, he explained that he had been suffering from an infection in his foot for two years. Doctors couldn't get it under control because the bacteria were resistant to all antibiotics. He was scheduled to have his foot operated on a fourth time the next day. Could I put him in touch with someone in Georgia? He was afraid that before long he would lose his foot.

More than any research I have done, his call hit me between the eyes. Never before had I been so aware of the power that bacteria continue to wield over us. We have grown up with the certainty that every bacterial infection can be cured by antibiotics. Most of us have no idea of the destruction that bacteria are capable of rendering, because our doctors prescribe drugs at the slightest symptom.

One year after I received this call, I accompanied an expedition of botanists and fragrance researchers to a rain forest in Madagascar. One night, as I slept in my hammock, I woke up, and my right foot was hot, red and swollen. The next morning I could hardly walk. Bacteria must have entered places where my sandals had rubbed against my skin while we were hiking. The doctor accompanying the expedition gave me some antibiotics that he found in his first-aid kit. The effect was hit or miss – more miss than hit, in fact. Four days later, I arrived home – with my foot still swollen. My GP prescribed some other antibiotics and luckily they worked. He cut to the chase: 'That could have been the end of you.'

At that point, however, I no longer needed that kind of graphic demonstration of the power of bacteria, since I had already started doing research for this book. The 80-year-old history of the tiny phages and their potential role in reining in the antibiotic resistance crisis were constantly on my mind.

The fascination produced by phage therapy is particularly striking as I write these lines. In Southeast Asia, veterinarians and doctors are combating bird flu. A pandemic is in the making. This was only just averted in the case of SARS, a new atypical kind of pneumonia. These health crises show viruses in their familiar role – as lethal villains. Phage therapy takes this image and turns it upside down, turning the bad guys into unexpected allies.

This book is not a health manual whose purpose is to testify to the efficacy of phages. First, it's too early to reach a clear conclusion about their effectiveness. Researchers are still working on this. Second, I found the detective work on the origins of the captivating idea that bacteria can be fought with their natural enemies at least as interesting as the analysis of phages' curative powers. I hope that this has led to a book that sheds some light on the sometimes winding paths of medical research and in turn provides some insight into an area of our society that is becoming increasingly significant. Never has so much medical research been undertaken as at the present time, nor has so much money ever been spent to cure us of diseases.

This English edition came about some three years after the German edition was published. I have taken great pains to update the material in the book. As I did so, I saw that some companies had been confronted with scientific or financial obstacles, leading them to abandon their projects altogether. On the other hand, other companies and university researchers have joined the ranks of phage therapy research, contributing good ideas. What they require is support from public and private sponsors in order to produce drugs from phages. They are desperately needed.

I could not have written this book without the help of many researchers, doctors, patients, librarians and helpers. They provided me with information, books and photos, gave me accommodation, told me about their lives, interpreted or handed out advice. I extend my gratitude to all of them.

I would specifically like to thank Elizabeth Kutter, Hans-Wolfgang Ackermann and Harald Brüssow for sharing their expertise. Zemphira Alavidze, Nino Chanishvili, Liana Gache-chiladze, David Gamrekeli and Mzia Kutateladze not only provided me with exhaustive information, but made my research in Georgia possible in the first place. I will never forget their hospitality.

Reto Schneider, Elizabeth Kutter and my wife Susanne read the entire manuscript. I thank them for their countless suggestions for improvement in style and content. I also express my thanks to my translator Karen Leube, my editors Sara Abdulla (Macmillan) and Wolfgang Gartmann (Piper), and the team at Aardvark Editorial. Without the support of Tamedia AG, the publisher of *Facts* news magazine, this book would not have been possible. *Facts*, my employer, continued my salary while I worked on this book, and Tamedia's media forum paid for the research expenses. I would like to thank my colleagues at *Facts*, Odette Frey, Beate Kittl and Rainer Klose, for their willingness to put up with the additional work and reorganization brought about by my absence. The English translation was generously funded by the Foundation for Fatal Rare Diseases. Thomas Fritschi and Rich Weber drew the graphics for Figures 3.4 and 3.5. I thank Susanne and Julia for putting up with a husband and father who was more of a phantom for a year and, at times, an overworked, nervous one at that.

Thomas Häusler

# 1

## at the limits of medicine

At some point during those fateful days, microbes barged their way into Alfred Gertler's life. They lodged themselves there, spread all over and took control. They devoured his bones.

Gertler, who was 41 at the time, had signed up for a gig on the cruise ship MS *Maasdam* for a few months. His lopsided house in Toronto was 'full of my two sons' diapers, but we were pretty short on money', as he put it. He hoped to remedy the situation by working as a musician on the ship. In March 1996, the *Maasdam* was cruising in the Pacific along the shores of Central America. Gertler left the luxury liner during a stop in Caldera, Costa Rica to rest from his hard work as a bass player. It was intended to be a break from the gruelling daily concerts, which ended with the players up to their ankles in music because there wasn't even enough time to turn pages. Yes, a hike in the hills at the end of the Pacific Ocean would do him good. That's not quite how it turned out. On the way back he lost sight of the footpath. Since the 4 o'clock rehearsal would begin shortly, Gertler decided to climb down the steep slope below the path down to the road. This decision would change his life forever.

It was a short fall, less than 15 ft. The root Gertler grabbed onto on his way down broke off. Suddenly he was lying on his back, completely unable to move. 'My foot was folded like a sock', he recalls. 'My bones were sticking out. They looked white and soft', he told me when I visited him in Toronto six years later. Some local souvenir vendors strapped him to a board and carried him down to the city, where his foot was bandaged and put in makeshift splints and his wounds rinsed out. And all this without any anaesthesia. Gertler was afraid that if he had an injection, his foot might become infected. He was transported over a bumpy road to the hospital in San José, the capital. Six days later, he flew back to Toronto. His GP inspected his cast and decided to leave it at that.

At this point, the bacteria must have already sneaked into the wounds. The germs would cripple his foot, confine him to bed for years and expose the limits of medicine. Perhaps the microbes were lurking in the soil where he tumbled down the slope. Or maybe they were hiding in the mildewed bathroom in the San José hospital, which had filled Gertler with such utter panic that he had scrubbed the shower for two hours before setting foot in it. There's no way to say for sure, since the bacteria silently ran rampant under the cast until the pain in Gertler's swelling foot drove him to the hospital. Too late.

The orthopaedist at Sunnybrook Hospital made a shattering diagnosis: 'If you don't die, then your foot will.' Gertler was given an ultimatum: have his foot amputated or the bacteria would continue to eat their way up the bones, damaging first the foot, then the lower leg and then the thigh. Gertler panicked. He didn't want to lose his foot. Desperate, he dragged himself to another hospital. The doctors there agreed that his foot should be amputated. Gertler again refused to have the operation. That was the beginning of a battle between the doctors and the staphylococci – the germs commonly known as 'staph' – in his ankle. They used the strongest medicine available. For more than two years, Gertler continuously received antibiotics. For a year, they were injected directly into his bloodstream via an electric pump.

These miracle drugs, which many people consider to be the greatest medical achievement of the 20th century, were a complete flop. The sustained attack of cloxacillin and ciprofloxacin, the drug made famous by the anthrax attacks in the US in autumn 2001, barely dented the microbes in Gertler's foot. They hunkered down in the bone, gnawed away at it, kept the joint swollen for years and held open two gaping, weeping wounds. Gertler spent most of the time in bed: 'Even going to the pharmacy around the corner on my crutches and hobbling back was enough to make all hell break loose.' The movement in the broken joint ejected the microbes from their hiding places in the bone and the neighbouring tissue and sent them to the bloodstream, where they multiplied like crazy. The ensuing blood poisoning and the overreaction of the immune system connected with it – doctors refer to this life-threatening combination as 'sepsis' – confined Gertler to bed for weeks with fever, chills and continuous fatigue. 'The worst thing was when the wounds healed up again', he told me, 'because then the hope returned that it was all over. I finally got out of bed, played music and, whoosh, it broke out again.' He experienced this so often that you could hardly hear the bitter disappointment in his voice any more.

The strange thing was that, in a test, the microbes didn't even show a particularly raised resistance to antibiotics. Antibiotic resistance is a nightmare for infectious disease specialists, who increasingly have to stand by and watch helplessly as bacteria dupe the drugs aimed at them and people die because no drugs work. Ironically, an estimated 10 per cent of all patients in hospitals are infected by bacteria. In the US alone that makes 2 million patients per year, and 90,000 of them die, 70 per cent from highly resistant bacteria.[1]

In Gertler's case, the antibiotics simply couldn't reach the colonies of staphylococci in the bones: 'I couldn't believe it. I had an infection with staph – one of the most common complications of surgery – and not a thing could be done about it.' Gertler was just one of many patients who suffer the same fate. Just because an antibiotic destroys the germs in a test tube doesn't guarantee that it will pack the same punch in a patient's body. The drugs are especially ineffective in places with a poor blood supply, like bones. Victims are constantly subjected to new operations to cut out affected tissue which disfigures them more and more.

## Mysterious hope

For Gertler, doing gigs as a bass player was now out of the question. Money became tight. And being deprived of music – his passion – was just as bad. In his old house on Kensington Market in Toronto, whose esoteric shops, jazz cafés and vegetarian restaurants are reminiscent of hippy capital Haight-Ashbury in San Francisco, there are always lots of different radios in the different rooms all tuned to the same jazz station, even at night. 'The infection was like a wish from hell. Hey – you want to sit in bed all day and listen to music? Okay, here it is.'

It got even worse. His wife moved out and took the two children with her. 'I don't blame her. It's pretty hard to stand it with a guy who lies in bed all day, is tired all the time and constantly complains that he's in pain', he told me as we sat sipping tea in his kitchen. The corners and walls of Gertler's house are full of pictures of his sons. On the doorframe you can still see the marks he used to record the boys' heights as they grew. The last is at 89 cm. When Gertler sees it, he always thinks of what the doctor told him: 'When your foot gets bigger than your life, get rid of it.'

Gertler's foot is still attached. He started fighting, looking for ways to conquer the bugs and liberate his foot from the destructive staph. Since Gertler wasn't internet savvy, his brothers and his parents helped him to search for information. As it turned out, it wasn't the Web that helped the desperate musician. It was pure coincidence. In early 2000, nearly four years after his fateful accident, a friend of Gertler's collapsed while riding his bike. The emergency doctors' diagnosis was acute blood poisoning – caused by staph.

Gertler paid his friend a visit in the hospital. The 6 February 2000 issue of the *New York Times Magazine* was lying on his tray table. In it was an article about an amazing treatment from the Republic of Georgia.[2] Gertler devoured the article, learning that in this distant land in the Caucasus, microbes were being used to fight microbes. The physicians there let special viruses loose on the bacteria. For each type of bacteria, nature provides a matching virus that decimates the pack with chilling efficiency, leaving humans unharmed.

The treatment operates on the principle: 'My enemy's enemy is my friend.' It was also used in the Western world well into the 1940s. Yet the success of the treatment varied widely and, with the discovery of penicillin, it was abandoned. In contrast, in the Soviet Union, therapy involving bacterio-phages, as the tiny bacteria-exterminators are called, continued over the years – almost unnoticed by Western medi-cine. In Gertler, the seeds of hope began to germinate. Especially intriguing was a bit in the feature describing how, in Georgia, the method was used to target cases in which antibi-otics didn't work. 'That same evening I limped off and made ten copies of the article.'

At the very end of the text, it was reported that a bacterio-phage conference was about to be held in Montreal, only 530 km from his home. Gertler contacted the organizer, Michael DuBow, a researcher at McGill University in Montreal. 'DuBow told me that it was a scientific conference, but I could attend if I paid the registration fee.' The conference was primarily dedicated to basic research on phages, which had also been carried out in the West. However, the growing crisis regarding resistance to antibiotics had sparked the interest of several biotech companies and scientists in phage therapy, so some lectures on this topic were scheduled too.

So here was Alfred Gertler, a jazz bassist, at his first scientific conference. Even among the crowd of scientists, who were otherwise oblivious to the outside world, he stuck out, with his crutches and his inevitable Greek fisherman's cap. 'In the breaks he hung around the smokers who would stand around outside on the steps', recalls Elizabeth Kutter, a phage researcher who later ended up helping Gertler. It didn't take long for him to get to know several scientists who were exploring phage therapy. Treating him, however, was something altogether different, too much of a risk for them. As a treatment that had not been sufficiently tested or approved, no one in Canada or the US was allowed to use it without risking serious consequences from the health authorities. 'Despite this, though, there were a few scientists who offered to help me right away', Gertler said. "Bring us a bacteria sample from your wound", they said, "and we'll pick out the phages that will work in our lab."'

On the final day of the conference, on a boat ride on the St Lawrence River, Gertler struck up a new acquaintance during a smoking break. 'I asked the guy if I could have a puff. I don't really smoke, but it restricts the circulation in my foot and helps me to get home without it hurting so much. And we got to talking.' The chain-smoking researcher was from Georgia, the country featured in the *New York Times Magazine* article.

He asked me for a bacteria sample. I hobbled to the restroom. A storm was in progress and rocked the boat like crazy. I had to be really careful not to fall over as I was taking off my shoe and sock. I took one of the cotton swabs that I'd been carrying around with me the whole time during the conference, scraped off some secretion from the wound and put my sock and shoe back on again. Then I limped back on deck.

The researcher gave Gertler his business card. It said: Revaz Adamia, Chairman of the Defence and Security Committee of the Georgian Parliament. 'I thought he was from the KGB, and I'd never hear back from him again.' Yet soon after the encounter, Adamia, who worked full time as the head of a lab at the Eliava Institute for Bacteriophages, Microbiology and Virology, sent Gertler an email from the Georgian capital of Tbilisi: 'We have the right phages. Treatment at the local hospital is no problem.'

## So close and yet so far away

But Gertler decided he would rather have his foot treated in Canada or the US. He was too uncertain about what to expect in Georgia, where a civil war had been raging only a short time before. That's the way he saw it back then, anyway. Gertler preferred to accept the offer of an Israeli researcher he had also met at the conference. The Israeli cultivated phages for diagnostic purposes and had found a virus in his inventory that, in a test tube, had wiped out the staph from Gertler's foot. The researcher planned to send him these phages; all Gertler had to do was to find a doctor who would treat him with them. He put together a file of articles and printouts from the internet with information about phage therapy and set out to find a Canadian doctor who would be willing to risk an attempt at the exotic method. 'I felt like a nut, the way I limped from practice to practice with my bundle of paper all marked up with yellow highlighter.' His limping only landed him in a blind alley. There wasn't a single doctor willing to put his or her head on the block.

Finally, Gertler happened upon some exciting news on the internet. In 1999, doctors at a hospital in Toronto – his own city! – had used phages to snatch a dying woman from the jaws of death. The woman had Marfan syndrome, a serious genetic disorder, and was also suffering from an acute staph infection, this time of the heart. The microbes were resistant to everything the doctors had pumped into her. Organ after organ had failed.

By chance, the woman's son had heard about a company called Phage Therapeutics in Seattle, Washington, which had just developed a bacteriophage drug used to treat highly resistant staphylococci. After my visit to Gertler, I talked to Richard Honour, then the head of the tiny start-up. Honour vividly recalled how a man had called him, desperate for help. 'Mom is dying!', he yelled into the telephone. Honour immediately agreed to provide the drug for emergency treatment, despite the fact that it had only been tested in animal trials up to that point, meaning that to all intents and purposes he was circumventing the law. The two swore each other to confidentiality. The doctors quickly had a bacteria sample flown to Phage Therapeutics.

Honour's researchers put the deadly staphylococci in a test tube and set their phages on them. 'And zoom, the phage killed the bacteria', Honour told me. He sent a few vials to the Canadian border, where they were received by a hospital worker. Following the instructions of Phage Therapeutics, the doctors sprayed the phage solution on the dying woman's heart, which had been exposed in the chest cavity. The next day, they injected a gigantic dose – 100 billion viruses – into her. Within 20 hours, the patient had recovered, and the bacteria had completely disappeared from her heart and her blood. According to Honour, she died of the heart problem caused by the genetic disorder several months later, but at the time of death, she was still free of the staph.

The pact of confidentiality that Honour and the hospital had agreed to was short-lived. The identity of the person responsible for leaking the information to the press remains unknown, but the news that a dreaded 'superbug' infection had been instantaneously cured made the headlines. The Canadian health authority threatened to have Honour arrested. The patient's family and her physicians went underground. For six months, Gertler tried to find out the names of the hospital or the doctors. Although various sources gave him tips, including email addresses, 'they refused to communicate, and I finally figured out that the doctors didn't want to be reached. I don't blame them. After all, they had already put their careers on the line once before.'

## Rusty surgical instruments and vodka as a disinfectant

Just then, Gertler got a call from Elizabeth Kutter, the phage researcher he had met at the conference who was in close contact with the scientists from Georgia. Kutter was planning another trip to Tbilisi and asked him if he'd like to join her. 'From the outset I had told him to get treatment in Tbilisi, because the doctors there had years of experience with phage therapy', Kutter says. Her call couldn't have come at a better moment. Gertler was ready to go to Georgia this time. In the three weeks leading up to the trip, his family arranged the flight, visa and money for Gertler, whose illness had taken a huge financial and physical toll.

At dawn on 29 January 2001, Kutter and Gertler arrived at Tbilisi airport. The long journey had stirred up the germs in

Gertler's foot and acute blood poisoning was imminent. After a brief rest at the flat of Liana Gachechiladze, a phage researcher from the Eliava Institute, his hosts took him to the tiny diagnostic practice, Diagnos 90. Gachechiladze, along with fellow researchers, had founded the practice as a stopgap after the Eliava Institute had been hit hard by the collapse of the Soviet Union. The mighty production units that had once churned out masses of phage drugs had been cannibalized, and the staff had shrunk from more than 1000 to fewer than 100. The employees who had stayed on in the crumbling institute had to supplement their $30-a-month salary that the impoverished government paid (or sometimes didn't) by taking on a second job, like the one Liana had at Diagnos 90.

The practice was situated in a former gatehouse at the entrance to the extensive institute complex. Stray dogs greeted Gertler as he hobbled up, and signs of fire on the walls only added to the desolate conditions of the building. Gertler, completely exhausted, must have felt queasy at the sight of the run-down place. As soon as he entered the practice, a physician's assistant wanted to take a sample in the freezing exam cubicle, and utter fear descended upon him. 'She went straight at my ankle with this rusty coat hanger.' Was everything going to get even worse here? Wouldn't it have been better to have his foot amputated after all?

Two days later, these fears gave way to cautious admiration, as Inga Georgadze, the head of Diagnos 90, presented the results of their analyses. She had tested a long list of antibiotics and phages for their effectiveness on the staph. Gertler, whose odyssey had turned him into an antibiotics specialist, saw many drugs on the list he had never heard of. In the ruins of a battered country, here were scientists and doctors who were doing their best to maintain their skills and their art – and to help him.

Several phages from the institute's collection completely eradicated his staph in the test. The Eliava researchers combined them into a powerful cocktail. An X-ray revealed the extent of the damage to his foot and confronted Gertler with the severity of his injury. 'The doctors squirted a contrast medium with a lot of pressure in the wound on the outer side of my ankle, and it came out through the opening on the inside of my ankle', he recalled, shuddering at the thought. 'Over the years, I had made a point of washing out the wounds every day, but I had no idea that they were so directly connected.'

After the weekend, Elizabeth Kutter and the Eliava researchers brought Gertler to the Tbilisi central hospital, where surgeon Guram Gvasalia, who had years of experience with phage therapy, wanted to perform the treatment. In the tight, windowless foyer of the 12-storey hospital, the only bit of light came from a red Coke machine. The gloomy lift was operated by a toothless elderly man, whose services proved to be necessary, since the tin box could only be opened by fiddling around with the inner workings of the clanking doors. As a special guest, Gertler was allowed to pick out a room, either one that was close to the toilet in the corridor or one with a view of the city. Otherwise, the features were the same as those of the other freezing hospital rooms: bare, peeling plaster, holes in the unfinished parquet and ancient iron pipes. Patients' families crouched outside in the dark corridor. They cooked meals for their sick relatives or went to the chemist to pick up prescriptions. The hospital received practically no money from the city, so the patients had to have their drugs bought for them.

For Gertler, a cook was hired, who 'bought vegetables, cooked like a dream and spoke to me in Georgian as if she were my mother'. One night he was awakened by a fellow patient's family member looking for a bed to sleep in. Although Gertler had become convinced of the medical skills of his Georgian friends, the strange and shabby surroundings began to take a toll on his mental state. Kutter had to get him several bottles of vodka. 'Alfred kept wiping off everything he saw with it to disinfect it', she recalls.

The treatment began. In the operating room, Gvasalia, the surgeon, carefully washed out the pierced foot. Then he flushed its deep caverns with the phage solution before he attached IV tubes to its two open sides. The tubes pumped into the injured joint the staph killers and enzymes to break

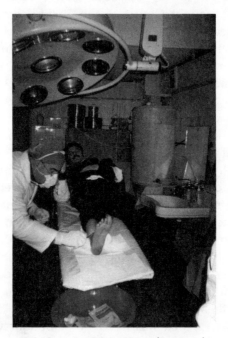

**Figure 1.1** Surgeon Guram Gvasalia uses phages to treat Alfred Gertler's foot at the Tbilisi central hospital

down the scarred tissue. The next day, the doctor repeated the procedure. In addition, he placed slivers of a biodegradable membrane – impregnated with phages and selected antibiotics – deep inside Gertler's foot. As the strips slowly disintegrated, new phages and drugs continuously forced their way into the tissue. The cornered microbes experienced sustained bombardment. After three days, the doctors could no longer find the staphylococci, previously teeming in the wound secretion.

Gertler had to hold out in the central hospital for another two weeks. Gvasalia urged him to have surgery to stabilize the joint that had been worn down by infection for so many years. There might still be nests of staphylococci deep inside the bone that would be released by the constant friction of walking.

Now, four-and-a-half years later, Alfred Gertler is doing better than at any time between his fateful fall and his stay in Tbilisi. Since undergoing phage therapy, he no longer takes antibiotics. He walks to concerts from his jazz-filled house, supporting himself on crutches and hobbling, carrying weights that the doctor has outlawed – and not once has he come down with acute blood poisoning. From a strictly scientific point of view, no one can really say how much of a role phage therapy has played in this improvement. In an isolated example like this, it could be a simple case of spontaneous healing, or some other factors may have had an impact. The only way to prove the effect of the bacteria killers is to carry out carefully planned and documented studies with a large number of patients.

## Wily enemies call for unorthodox methods

After 50 years in the cold, the massive increase in resistant bacteria has brought phage therapy back to the attention of researchers in the West. Humans are in a constant struggle against microbes that we will never win. Bacteria have always proved to be tricky. This is why we need new drugs, as well as the rediscovery of old ones, since new antibiotics are extremely slow to enter the market. Therefore, for several years now, a small number of universities and biotech companies have been working on phage drugs again.

Yet many infectious disease specialists continue to be extremely sceptical. They argue that researchers in the controversial early history of phage therapy never managed to produce clear proof of its efficacy. The critics are also suspicious of the work of researchers from the former Eastern bloc because most of their results were reported years ago in Russian or Georgian journals – making them almost impossible to check. Political turbulence and economic decline in the countries of the former Soviet Union only add to the doubts.

To get an idea of the contribution that phage therapy can make to the fight against bacteria, you have to go back to its beginnings around 1920 and comb through the reports that numerous scientists have amassed since. What you find is a rich history, filled with hopes and disappointments, eccentric heroes and tragic fates, gigantic experiments with thousands of subjects and evidence that our grandparents swallowed phage medicine. In France, you can even come across retired researchers who used phages around 1970 to cure patients who had been declared incurable. Now it's time to reactivate all this knowledge after years of fooling ourselves into thinking that modern medicine is the victor.

# 2

## invincible microbes

'It is time to close the book on infectious diseases', US Surgeon General William H. Stewart announced in 1969. 'The war against pestilence is over.'[1] According to America's highest ranking physician, scientists should have been investigating cancer instead of tuberculosis and replacing the study of cholera with research on heart attacks. Stewart wasn't the only one who was so optimistic. The development of penicillin in the early 1940s in particular had brought euphoria. The medical community was sure that it was just a matter of time before bacteria would be conquered. They were wrong.

### Reconquest

Infectious diseases had only taken a breather. Take tuberculosis. In several countries of the former Soviet Union, the number of people infected with tuberculosis doubled within just seven years of the collapse of the USSR. Today, in the area around the Aral Sea, 300 out of every 100,000 people suffer from TB. In the prisons, the number of infected people is up to 100 times higher.[2] On a global level, some 2 billion people carry the tuberculosis bacterium, approximately one-third of the world's population. It is estimated that 2 million people die of tuberculosis every year.[3] In England and Wales, the number of infections rose by 20 per cent between 1994 and 2004.[4] In countries of the former Eastern bloc in particular, many tuberculosis bugs are already multi-resistant – the regular drugs are no longer effective. Yet the hardy microbes don't stay put; they've also turned up in places such as Chicago and London. In the optimistic Stewart's home country, New York City to be exact, there was even an epidemic of multi-drug-resistant tuberculosis from 1972 to 1992. In 1992 alone, doctors registered 441 new infections. Many of the patients died, despite being involuntarily admitted for treatment in quarantine on Roosevelt Island. Quashing the epidemic cost about a billion dollars.[5]

The catastrophe demonstrated what had been simmering away unnoticed by the public. Antibiotics, the magic swords used to kill all kinds of bacteria, were becoming increasingly dull. Microbes were defending themselves and continue to do so by pulling ever newer tricks out of their hats. Every microbe that makes people sick has armed itself to some extent in the fight against antibiotics. Most of them are horrifyingly successful. When penicillin was introduced, it did an excellent job of stamping out bacteria from the *Staphylococcus aureus* species. These days, more than 95 per cent of these bacteria, which cause boils, blood poisoning and bone inflammation, are resistant to penicillin. In several parts of the world, 98 per cent of infections with *Neisseria gonorrhoeae*, the cause of gonorrhoea, can't be cured by this antibiotic. There are already microbes that checkmate every single one of the more than one hundred antibiotics in existence.

Today, *Staphylococcus aureus* alone kills 1400 people in England and Wales every year. In 1000 of these cases, the bacteria are multi-resistant.[6] In 2001, 90,000 people in the US died of an infection. Ten years earlier, the number was only 15,000.[7] The World Health Organization (WHO) warns that 'drug resistance threatens to reverse medical progress'. Harmless illnesses like tonsillitis or ear infections may be on the verge of becoming incurable again. In 1996, Hiroshi Nakajima,

WHO director-general at the time, warned: 'We are standing on the brink of a global crisis in infectious diseases. No country is safe from them. No country can any longer afford to ignore their threat.'[8]

## 'Thanks to PENICILLIN ... He Will Come Home'

With these worrisome developments in mind, Surgeon General Stewart and his contemporaries' euphoria seems rather naive. Yet a closer look at the situation prior to the discovery of penicillin may explain their optimism. The people who were alive in 1969, who had experienced the miracle of penicillin first hand, knew only too well what it was like before the wonder drug appeared on the scene. A slip of the shears while trimming the rose bushes could result in fatal blood poisoning. Nearly everyone's grandparents had lost siblings to diphtheria. A hospital in the 1930s was clogged with patients fighting pneumonia, blood poisoning or tuberculosis, with little or no chance of winning the battle.

Country doctor Jean-Pierre Feihl had a practice in the small town of Moudon in western Switzerland for many years. He remembers this time well: 'It was a tragedy', says the 85-year-old retired physician, 'especially when I think of the young people who had a bone infection caused by staph. You don't hear about cases like that any more, but back then it was a frequent occurrence. The patients suffered from pain in their bones and fever for years on end. They became emaciated and were often fatigued.' It's true that the improved hygiene and some vaccines kept many epidemics in check, and antiserums and sulphonamides, which were introduced in 1935, cured certain infectious diseases. However, in 1938, over 10 per cent of people in the US still died of pneumonia or tuberculosis. In England, microbes led the list of causes of death as well.[9]

The appearance of penicillin was a marvel. It proved to be extremely effective against staphylococci, streptococci, which cause pneumonia, and against diphtheria bacteria and many others. The death rate of pneumonia dropped from 30 per cent to 6 per cent. The drug was industrially produced in America for the first time midway into the Second World War and was reserved for use by the army until after D-Day on 6 June 1944, when it was made available to civilians as well. Penicillin was immediately on sale in pharmacies without a prescription, and adverts proclaimed its miraculous healing properties. In 1944, an advert sponsored by Schenley Laboratories showed a picture of an injured GI, with the words 'Thanks to PENICILLIN ... He Will Come Home'.

## An unheeded warning

Alexander Fleming actually discovered penicillin in 1928 but shelved further research on the drug because of technical difficulties. During the war, he quickly became a star as 'the greatest scientist of the 20th century' and appeared on the cover of *Time*. In 1945, Fleming was awarded the Nobel Prize along with Howard Florey and Ernst Boris Chain, who had stumbled across his discovery while perusing old medical journals in the late 1930s, and catapulted penicillin from the lab to the hospital.

From the very start, however, Fleming feared that the celebrated drug could cease to be effective if it were used at random. He concluded his Nobel lecture with a clear warning: bacteria can easily be educated to become resistant. Careless treatment with an underdosage of penicillin is enough, he cautioned. In this speech, Fleming already prophesied that deaths would be caused by resistant bacteria. The British scientist based his counsel on experiments he had carried out himself. He had exposed bacteria to ever higher penicillin concentrations and had seen some cells survive and proliferate. In 1945, in an interview with the *New York Times*, he said: 'There is probably no chemo-therapeutic drug to which, in suitable circumstances, the bacteria cannot react by in some way acquiring "fastness".'[10]

It wasn't long before Fleming's fear became reality. As early as 1944, some patients could not be cured by penicillin. Two years later, a London hospital announced that the antibiotic was ineffective against 14 per cent of staphylococci. By 1949, this number had jumped to 59 per cent, and England was not alone. In an article published in the *Schweizer Apotheken-zeitung* (Swiss Pharmacy Journal) in 1955, it was reported that 75–80 per cent of all staphylococci were resistant to penicillin, and the stubborn bug was running rampant in other countries.[11] Physicians and health authorities were not particularly concerned. Directly after introducing penicillin, the pharmaceutical companies had quickly fielded a slew of new antibiotics. In 1943, microbiologist Selman Waksman discovered streptomycin, which attacked the tubercule microbe that did not react to penicillin. In 1947, chloramphenicol, a wide-spectrum antibiotic, was launched. More followed. The chemical sledgehammer appeared to be unbeatable. Around 1955, several countries classified antibiotics as prescription drugs, halting the most extreme cases of improper use.

Yet the microbes continued to defend themselves. In the 1970s, penicillin-resistant *Neisseria* turned up all over the world. It is fairly certain that these can be traced back to brothels in Southeast Asia. During the US occupation, prostitutes there were given penicillin as a preventive measure, which in turn also protected their military visitors. Today, many countries all over the world are struggling with this offspring of the Vietnam War. In Southeast Asia, 98 per cent of *Neisseria* are resistant – not only to penicillin but also to many other drugs.

## Natural-born killers

The arms race between microbes and man is an inevitable result of evolution. Every new antibiotic that we use leads to the selection of hardy bugs within a short time. Antibiotics expert Stuart Levy of Tufts University, Boston calls it the 'antibiotic paradox'.[12] If a microbe colony is attacked by a drug, a few cells will often survive – because their ancestors picked up a gene that conferred resistance or due to random, protective, genetic mutations. These plucky cells then thrive because the drug kills off all their competitors for food, opening up the field to them. Under optimal conditions, certain microbes divide every 30 minutes. Theoretically, in 24 hours, over 200,000 billion cells can emerge from a single cell. That's how a drug-resistant microbial strain is born.

Bacteria do not actively pursue this armament. For example, they don't deliberately mutate with the intent of surviving an antibiotic attack. Mutations come about by chance. Some of them make the cell sick, some kill it, others have no effect at all and still others bestow upon their carriers a resistance to a certain antibiotic. This protective alteration can happen before or during antibiotic treatment.

## Human breeding machines

Sometimes a fatal spiral is set off inside one patient. New bug variants keep one-upping whatever antibiotic the doctors inject. The patient becomes an incubator for a new super-race of bacteria. At the Milan congress of European infectious disease specialists in 2002, Ian Phillips of St Thomas' Hospital in London described this kind of case.[13] Shortly after admitting a patient, doctors diagnosed a bone infection brought about by streptococci, a scary diagnosis because many drugs barely reach the bones due to poor circulation. It is easy for the under-dosage described by Fleming to occur in such cases. Phillips and his colleagues started treatment with penicillin – with no effect. The same was true for clindamycin, cloxacillin and a whole list of antibiotics with other exotic names – ciprofloxacin, rifampicin in combination with ciprofloxacin – no effect. At each round a few of the barricaded microbes survived and multiplied again. After months of failed attempts, a break-through finally came: vancomycin worked. It was a close call.

The resistance career of *Staphylococcus aureus* provides the opportunity for a case study of the agility that bacteria show in the race to survive. As already mentioned, just a few years after penicillin was introduced, 15 per cent of staphylococci were resistant in some places. Ten years later, the number had soared to over 70 per cent, and today it is 95 per cent. To head off the prospect that staph infections could be completely untreatable, chemists developed methicillin, a synthetically modified variant of natural penicillin, produced from a fungus.

In 1960, methicillin was used to treat staphylococci for the first time. A year later the first resistant bacteria appeared, starting the race all over again. Doctors kept attacking the microbes with new antibiotics – only to stand by and watch as the bacteria fought back. Chloramphenicol, ciprofloxacin, clin-damycin, erythromycin, gentamicin, imipenem, tetracycline, trimethoprim – the list of drugs that some staphylococci had armed themselves against became increasingly longer. The abbreviation MRSA was originally the label for methicillin-resistant *Staphylococcus aureus*. It now refers to multi-resistant *Staphylococcus*.

Recently, the first cases occurred in the US and Japan in which vancomycin, often the drug of last resort, was ineffec-tive. In many countries, well over 40 per cent of the bugs in patients are now multi-resistant. At the Congress of Clinical

Microbiology and Infectious Diseases in Milan, Helen Giamarellou, of the University of Athens Medical School, reported on the conditions at an orthopaedic hospital in Athens: 'Every day five new cases arrive with bone infections, and they all have MRSA. Tell me what we're supposed to do with these patients? What should we tell them?'

## Lazy doctors fuel the arms race

Doctors must also take some blame for this state of affairs. 'We set ourselves up for the current situation by excessively prescribing wide-spectrum antibiotics', says Andreas Widmer, hospital hygiene expert at the University Hospital Basel in Switzerland. The newer broad-spectrum agents destroy a wide variety of microbes at once. Preparations with a narrower spectrum, generally older ones, only attack a limited group of bacteria. For example, penicillin G is primarily active against so-called Gram-positive bacteria such as staphylococci and streptococci. Their name is derived from a staining method, 'Gram's stain', that pathologists use to make the cells visible under a microscope. Gram-positive bacteria have a simple cell wall, absorb Gram's stain easily and stain violet. Gram-negative bacteria, which include those that cause gonorrhoea, have a triple-layer cell wall, and do not retain the stain. They appear red. To treat both groups at the same time, the industry developed wide-spectrum antibiotics – at the urging of doctors. They spare the physician the trouble of making an exact diagnosis and save one or two days of waiting until the infection has been identified.

'Until recently many doctors immediately prescribed antibiotics when the patient had a fever', Widmer says. 'They administered it like aspirin. In many European countries this continues to be the case.' The result is that in many regions of the world, bacteria such as pneumococci, the culprits behind many cases of pneumonia, inner ear infection and meningitis, rarely respond to penicillin, which is actually still the optimal drug. A study of healthy children living in two communities in Utah shows how the use of antibiotics is driving up the rate of resistance. Overall, 10 per cent of the children had multi-resistant pneumococci in their noses. In the community where doctors were more circumspect in prescribing antibiotics, far fewer children were affected with the annoying bacteria than in the one where physicians were more liberal in doling out drugs.[14]

## Uninformed patients breed a menagerie of resistant microbes

Often the antibiotics prescribed by GPs are unnecessary and the only effect is the acceleration of the resistance spiral. This was shown by William Holmes, a GP from Nottingham. He is familiar with the problems of a daily practice: 'Most patients who go to a GP with a respiratory infection have one thing in mind: they want antibiotics.' In a study Holmes carried out involving 76 GPs and 787 patients, 77 per cent of patients requested antibiotics. The problem is that many respiratory infections are caused by viruses against which antibiotics help as much as Vitamin C against a headache. 'These patients aren't easily convinced otherwise, even if you use scientific evidence', says Holmes. 'They want antibiotics anyway.' The second problem is many doctors give in to the pressure. Of 581 patients who were prescribed an antibiotic, only one-fifth should 'definitely' have been given one, and some 150 'definitely not'. Most doctors said they went against their better judgement because of pressure from patients.

Depending on the assumed necessity for antibiotics, the doctors in Holmes' study made fine distinctions regarding the drugs they prescribed. Patients who they felt did not need antibiotics were usually given a prescription for tetracycline. When doctors wanted to effect a cure, they prescribed newer preparations. Both antibiotics contribute to resistance – including tetracycline that is prescribed as a placebo.[15]

Every person is colonized by billions and billions of completely diverse species of bacteria, which live on the skin, in the mouth, the nose and the intestines. Most are harmless or even beneficial, because they occupy the space that could otherwise be invaded by illness-causing bacteria. Once in a while, healthy people also harbour bacteria known as pathogens, for instance *Staphylococcus aureus,* in their noses. Why such species make some people sick and not others is not completely understood. The state of health of the person confronted with a bug is important. An intact immune system has no trouble stopping staphylococci moving from the mucous membrane of the nose to the inside of the body, but for one in poor shape, it isn't so easy.

When we take antibiotics – whether we abuse them or take them according to prescription – we not only kill harmful bacteria, but we destroy beneficial ones, too. Resistant bugs survive. They multiply and are passed on to other people via body contact, or pass their resistance genes to newcomers on the current host body. An entire resistant menagerie is created on us that doesn't appear to have any negative consequences at the moment – until our immune system is weakened by an illness or with advancing years, and the resistant bugs suddenly attack us.

Even children can collect a frightening group of beasts on their bodies. In 1991 and 1992, Michael Millar of the Royal Trust Hospital in London investigated the oral flora of healthy seven-to-eight-year-olds. In 37 per cent of the children, he found staphylococci, 5 per cent of which showed resistances.

In addition, in 72 per cent, he discovered *Haemophilus* bacteria, 30 per cent of which were armed against some type of antibiotic. It was a shattering discovery. *Haemophilus influenzae* is one cause of meningitis in children. In the US, resistant bugs from this species have already caused fatalities. 'This may be just the beginning of something that will take off in a big way', Millar told the *New Scientist* journal.[16] How true. A recent investigation showed that in 2001, 19 times more children in England and Wales had become infected with MRSA than 11 years earlier.[17]

## Dangerous hospitals

The dangerous seed cultivated by the constant use of antibiotics sprouts in the very places where we seek protection and healing: in hospitals and old people's homes. This is where infectious disease specialists are fighting the nastiest problems, because it's where most antibiotics are used and the sickest people live. Between 5 and 10 per cent of all patients treated in a hospital are infected there, often with resistant bugs. In the US, an estimated 2 million people are affected per year, 90,000 of whom die. In the UK there are 300,000 infected patients, with 5000 deaths. This oft-quoted number of deaths was estimated in the 1980s and was imprecise even then, however. Up-to-date and better quality information is not yet available.[18]

A whole host of microbes armed to the teeth lies in wait for unsuspecting patients. At the top of doctors' list of worries are the familiar multi-resistant staphylococci (MRSA), closely followed by enterococci that outwit vancomycin, the drug of last resort, earning them their own abbreviation, VRE (vancomycin-resistant enterococci). Then comes a series of Gram-negative bacteria such as *Klebsiella pneumoniae,*

*Pseudomonas aeruginosa* and *Acinetobacter baumannii*. The distribution of these disastrous microbes varies. In 2000, in the US, over half of all staph infections in hospitals were MRSA, while in England the rate was 39 per cent in 2004. Since 1991, the number of blood infections caused by MRSA has increased 50 times there. Not many countries have higher MRSA rates. On the other hand, the Netherlands, which has always done a great deal to keep the resistance plague in check, was able to keep the rate down to approximately 1 per cent.[19]

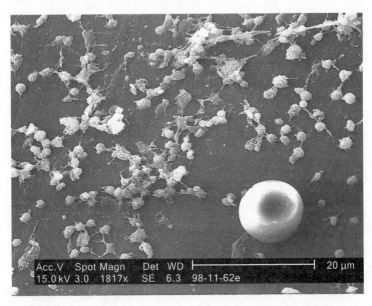

**Figure 2.1** *Staphylococcus aureus* on the luminal surface of an indwelling catheter, as seen through a scanning electron microscope. A red blood cell is also present with its biconcave shape. The sticky-looking substance woven between the round cocci bacteria is a so-called biofilm. It protects the bacteria that secrete it from attacks by antimicrobial agents

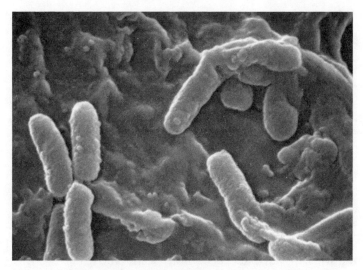

**Figure 2.2** *Pseudomonas aeruginosa,* as seen through a scanning electron microscope

MRSA, VRE and other opportunists use an environment that couldn't be more ideal. Hospitals, especially their intensive care units, are full of possibilities for hungry bugs. Patients with AIDS are often infected by microbes that don't affect healthy people but which exploit the opportunity provided by a depressed defence. Modern medicine uses transplants or cancer treatments during which the immune system is weakened for weeks at a time. 'Patients in intensive care units have a particularly high risk of becoming infected with resistant bacteria', says bacteriologist Hugh Pennington, current president of the British Society for General Microbiology. 'They are ventilated, have intravenous lines and a catheter in their bladder.' Added to that are surgical wounds that many patients have.

These are all gateways for the invasion of bacteria. Kleb-siellae have a penchant for moist ventilation tubes, while MRSA or pseudomonades enjoy jumping into wounds. Staphylococci reside in many people's noses without both-ering them. When there are multi-resistant variants, a fateful migration is triggered when patients touch their noses and then touch a nurse's hands. This has led to rigorous quarantine programmes in some hospitals. It may mean that a patient who is in hospital for a routine appendectomy is suddenly placed in isolation because hygienists have discovered MRSA in his nose. The nurse will only go into his room if she is wearing an apron, gloves and a surgical mask. The patient is not allowed to use the hospital cafeteria. It's solitary confine-ment. Quarantine is not lifted until doctors have eradicated the MRSA, which succeeds only occasionally. For the most part, in the past two decades, British hospitals decided not to pursue this search and destroy approach that countries such as the Netherlands rigorously enforced. 'That's why we have a problem now', says Pennington. A return to this strict regime is being discussed, but according to Pennington, many NHS hospitals, which are often filled to capacity, do not have the necessary facilities.

Even hospitals that have imposed strict quarantine aren't immune to outbreaks of infection that have terrible conse-quences. At the congress of European infectious disease specialists in Milan, an American doctor reported how vancomycin-resistant enterococci hit a kidney transplant department. The epidemic started with a patient who was hosting resistant enterococci. In healthy people, the oppor-tunistic bug is notorious for being a 'wimp' and lives in the intestine. In severely ill patients, however, it can run amok when it advances into the blood. Despite the carrier being isolated, doctors watched helplessly as the microbes attacked six victims one by one, each of whom was confined to his or her own isolation unit. Because physicians had to suppress the patients' immune system to prevent the freshly transplanted organ being rejected, the patients were gravely ill. The epidemic only stopped when the original carrier left the ward – because he had died.

Despite these dangers, hygiene is inadequate in many hospitals. Several studies have revealed that only half the staff members wash their hands thoroughly enough. 'Significant numbers of staff do not use hand hygiene in the way people would like it to be used', says Pennington. Disbelief among specialists prompted a study by the University of Munich. In 30 randomly selected practices and 25 hospitals, the cleanliness of endoscopes used for examining the digestive system was investigated. Over half the disinfected instruments were contaminated by bacteria. Even after participants were made aware of this, a second check revealed a contamination rate of 40 per cent.[20]

As every newspaper reader in the UK knows, people are worried about hospital conditions: 'My husband was in an NHS hospital for six months from 1999 to 2000. Visiting each day I was afraid that the very bad standards of cleanliness and hygiene would result in his suffering an infection on top of heart disease and a stroke', wrote a pensioner in a letter to the editor of the *British Medical Journal*.[21] No wonder, when there are hospitals in the country where the MRSA rate is over 90 per cent. In June 2005, when 2000 Britons were asked which areas the NHS should focus its spending on, the most frequent response was hygiene in hospitals.[22] Seventy-two-year-old Blanche Beynon of Pentyrch, Wales was so worried by past experiences that she travelled to Belgium for a knee operation instead of having the procedure done at the hospital in nearby Cardiff. Her husband had been infected with MRSA in this hospital prior to his death. Blanche Beynon herself had been infected by MRSA twice at Cardiff's Llandough Hospital. So Beynon preferred to pay £7500 for the operation abroad because the hospital couldn't guarantee that she wouldn't be infected a third time. 'I have 17 grandchildren and 17 great-grandchildren and wanted to live a bit longer to enjoy being with them', Beynon told the BBC.[23]

Politicians have taken these fears on board, demonstrated by the squabbling between the Labour Party and the Conservatives in the UK parliamentary elections in 2005. John Reid, then minister of health, accused the former Conservative government of neglecting to fight MRSA when the epidemic took off in the early 1990s. Michael Howard, leader of the opposition Conservative Party whose mother-in-law died of MRSA, turned the tables and attacked the Labour Party because its government had not done enough to promote hygiene in hospitals. An article in *The Lancet*, 'MRSA: how politicians are missing the point', scolded politicians for the unproductive quarrel. What was needed, the journal said, was not 'tit-for-tat political posturing' but rather the search and destroy approach as practised in the Netherlands.[24] In the meantime, several programmes have been launched in Britain aiming to improve hospital hygiene.[25] Cases of MRSA are more precisely recorded than in any other country, and individual NHS trusts must publish their MRSA statistics.

## 'Hans, you got the staph from us'

The consequences of this negligence are borne by people who are sometimes infected for years. And even if victims do not always die from the stubborn bugs, they are constantly plagued by unemployment, pain, amputations and the fear that at some point nothing will help.

For many, it all starts as it did for the German Hans Friedrich (not his real name) – with an accident. One Friday in 1997, the market researcher was repairing a light fixture in his basement. Absorbed in his work, he didn't notice the ladder slowly inching across the floor. It finally slid out from under him and he fell to the floor. He was immobile for 45 minutes until someone answered his calls for help. Racked with pain, he could barely move. At the hospital, doctors said his heel had been smashed. They cut open the heel bone and stabilized it with a metal plate – a routine procedure.

At least it was until the staph turned up. The bugs moved into the wound, gnawed away at the bone and flowed into the blood, giving Friedrich two miserable weeks. 'I felt so horrible that I couldn't eat', he says. Friedrich had no idea how he had picked up the microbes or where they had originated. After constantly nagging the staff, his question was finally answered by the surgeon who had done the procedure. 'He didn't do that until I told him I was a doctor', says Friedrich. 'Then he said, "Hans, you got the staph from us."'

Friedrich realized he wouldn't get any help from his surgeon: 'If I had continued to go to him for treatment, I would have lost my foot. I'm a market researcher by profession, and I wasn't able to turn off those mechanisms as I searched for a solution to my problem.' He carried out surveys and made calls all over Germany and Austria to find the best specialists. So far, no one has been able to offer a solution to his problem. After five operations and a long checklist of antibiotics that failed to help, Friedrich continues to live with an open wound on his foot and huge respect for his staph. 'It generally takes about four to six weeks for the guys to beat the new drug. They have even become resistant to some disinfectants, which means that I have to keep switching to new ones – that's dramatic.' Luckily there are periods when the bugs are quiet.

On his tour of practices in the German-speaking region of Europe, Friedrich met a large number of fellow sufferers. 'Many of them were disfigured because they had been cut open so often.' The record holder was a man from South Tyrol, who Friedrich met in a hospital in Innsbruck. 'This man had suffered 29 operations until he was finally liberated from his infection.' Friedrich himself has one more antibiotic option available to him – Zyvox, the newest kid on the pharmaceutical block. '20 tablets cost 1800 euros [over $2000]', says Friedrich. 'But I'm saving that option in case the staph in my foot becomes unbearable again.'

## 'Apocalypse now?'

That's a good idea. In May 1996, doctors in a Tokyo hospital operated on a four-month-old baby to correct a heart and lung defect. Two weeks later he came down with a fever and pus started oozing out of the incision. The culprits were MRSA. The doctors administered vancomycin, usually the life preserver in such cases. After 29 days of treatment, though, pus continued to ooze from the wound. The physicians supplemented the vancomycin with another antibiotic and, for a short time, the symptoms stopped. Twelve days later, the baby had another bout of high fever, and an abscess had formed under the wound. Finally, doctors opened up the baby's chest and cut out as much of the infected tissue as they could. In addition, for 17 days they pumped an unusual combination of antibiotics into the baby. He survived. 'I was extremely shocked, because this infection was just hideous. The patient suffered greatly', surgeon Keiichi Hiramatsu told a journalist from the BBC.[26]

The real shock for the doctors, however, was the failure of vancomycin. Since the multi-resistant staph had started spreading, this antibiotic had been the weapon of choice and now it had failed for the first time. It's true that the staph from the baby's chest had not proved to be completely resistant to the drug in the lab test. At a high dosage, it was still effective. But this was small comfort, because in cases where the microbes become lodged in a region of the body with poor circulation, the effective ingredients only achieve insufficient concentrations, meaning that the microbes are virtually resistant.

This incident was a sign of things to come. If it had happened once, it could happen again. And it did. In July 1997, the next case of vancomycin-resistant staph occurred in the US, and a month later another was recorded. As of today, around a dozen cases have been confirmed worldwide, and some of them have ended in death. Suddenly it seemed possible for a simple tonsil-lectomy to be a life-threatening operation if it were performed in a hospital overrun by killer bugs. 'Apocalypse now?' asked the title of an article in *The Lancet*.

What happened in July 2002 was even worse. The experts' nightmare came true: multi-resistant staph had acquired the vancomycin-resistance gene from enterococci. The result was a super-microbe that defies even the highest concentrations of the antibiotic. In 1992, a British researcher had observed this fatal gene transfer between enterococci and staphylococci in the laboratory. As a precaution, he destroyed the ultra-resistant microbe after the experiment. Ten years later it created itself.[27]

Vancomycin has been on the market since 1958. Linezolid, which the US Food and Drug Administration (FDA) trium-phantly approved in April 2000 especially to treat multi-resistant infections, only managed to gain the upper hand for a while. Exactly a year later, in April 2001, US infectious disease specialists announced the first five cases of enterococci that were resistant to linezolid. Three months later, a team from Boston had more bad news: the even more feared multi-resistant staph had pulled off the same trick. Doctors at Boston General had diagnosed an 85-year-old dialysis patient with peritonitis. The guilty bug was MRSA. Because the patient was allergic to vancomycin, they administered the brand-new linezolid. He was given the expensive drug for a month, but the infection persisted. When the lab test was repeated, it revealed the worst-case scenario. The bacteria had become resistant during the administration of linezolid. At that point, doctors pumped a desperate combination of six other antibiotics into the man, including synercid, which had also just received FDA approval. The bacteria did decrease. But three weeks later, the patient died of his original illness. In view of the impact the incident could have, the doctors' report on the case was guarded. They concluded their bad news in *The Lancet* as follows: 'The emergence of resistance to linezolid in MRSA is an unwelcome development.' Market researcher Friedrich would agree. The trade name of the remedy he had placed in his cache is Zyvox – whose effective ingredient is linezolid.[28]

## 'We just have to wait'

If you think you can avoid the obstreperous bacteria as long as you don't set foot in a hospital, you're mistaken. For several years now, doctors in some parts of the world have announced more and more cases in which people are infected with resistant bacteria outside hospitals, especially the old familiar MRSA. Doctors are worried because most of these patients do not follow any of the known risk patterns. They are young, healthy prior to contracting the infection and their immune systems appear to function normally. Many of the infected patients are in prison and others are athletes. Outbreaks involving these kinds of patients were reported in a fencing club in Colorado and among football teams in Indiana and Los Angeles County. The patients generally suffered from painful abscesses. Occasionally the bugs settled in the soft tissue or even in the bones or blood. Researchers are surprised that the staph was even able to spread among athletes who were not involved in contact sports, like the fencers. Apparently it only takes a shared towel for the bacteria to work their way from one victim to the next.[29]

When an international team of researchers analysed the bugs from outbreaks in the US, England and Australia, they made a surprising discovery: they were dealing with a very old acquaintance. The *Staphylococcus aureus* variant responsible for the outbreaks is a descendent of a strain that had struck hospitals during the 1950s. It was known as 80/81 then and had troubled doctors with its resistance to penicillin. At some point, 80/81 disappeared from hospitals – for reasons that remain unknown – only to turn up in the wild 50 years later, enhanced by a resistance to methicillin.[30]

The re-emerging 80/81 strain had increased its defences during its period of latency. Somewhere it picked up a gene that allowed it to pierce defence cells of the immune system. This makes these staphylococci quite aggressive. In addition to the skin, they also colonize the lung and lead to inflammations that doctors have trouble controlling. Within a short time, pus collects in the lung and parts of it are destroyed. For several children in France, nothing could be done.[31] In the US, a similarly frightening variant has recently come onto the scene. It doesn't eat its way through the lung, but gnaws through the skin instead. Such 'flesh-eating bugs' are continually making the headlines, but so far the culprits behind this so-called 'necrotizing fasciitis' have been members of other species of bacteria, *Streptococcus* or *Clostridium* for example. When this new, flesh-eating staph attacks, its multi-resistance forces doctors to reach immediately for exotic antibiotics. The bacteria are capable of gnawing through up to 12 cm of subcutaneous tissue per hour. The surgeons often have to cut out large pieces of tissue in order to keep the patient alive. Many patients require skin grafts and plastic surgery afterwards to eliminate traces of the damage.[32] 'This is not a big problem in the UK, yet', says Hugh Pennington. 'There were one or two incidences, but the organism is here.' And all the experts predict that it will become more frequent. According to Pennington, there are few precautionary measures that can be taken. '*Staphylococcus aureus* is very versatile. We just have to wait.'

## A fateful large-scale experiment

As tragic as it is when dangerous new bacterial strains develop or a new antibiotic is quickly met with a defence, it is inevitable. Strict hospital hygiene and a better use of antibiotics can help doctors to slow the spiral of resistance. But they are incapable of stopping it. That's just how evolution wants it. If we can't win in the arms race against microbes, the best strategy seems to be to just provoke them when it is absolutely necessary. At least that's what one would think.

In the early 1950s, the agricultural industry looked for additives to accelerate the growth rate of fattening animals. By coincidence, some US researchers happened across the fact that a gloop consisting of micro-organisms that was a byproduct of antibiotic manufacturing spurred the growth of chickens. Scientists first suspected that some kind of vitamin was behind the phenomenon. It turned out to be small amounts of the antibiotic chlortetracycline, which remained in the residue of the microbes after extraction. Soon, additional tests showed that the magical effect could also be achieved by adding small amounts of other antibiotics.

A new industry was born, and a huge long-term experiment was launched. Even today, no one knows exactly why animals grow more quickly as a result of this obscure treatment. Still, until recently, a large percentage of Europe and North America's antibiotic production was not used to treat sick people, but instead landed in the stomachs of healthy animals. In 1997, patients in the European Union (EU) swallowed 5500 tons of antibiotics. That same year, 5000 tons ended up in farmers' barns. Nearly 1600 tons were consumed as so-called 'growth promoters' and the rest were drugs.[33]

In 1969, a British committee of scientists had already issued a warning about growth promoters in the Swann Report. They pointed out that antibiotic doping would beef up resistant bugs in fattening animals. These microbes could leap over to people either directly or by way of meat, and perhaps trigger infections that would be difficult to cure. In response to this, some countries prohibited the use of substances that were also used in human medicine.

This step wasn't enough. Even the practice of using antibiotics that will only be consumed by animals can have dire consequences. This has been shown in the case of the animal antibiotic avoparcin. In Europe, this substance was fed to livestock for years. Unfortunately, avoparcin has a chemical make-up that is similar to the human antibiotic vancomycin. Thus, microbes that defend themselves against avoparcin are also resistant to vancomycin. For a long time, this didn't appear to be a problem. Back then, doctors only administered vancomycin in a few instances, and there were other alternatives for treating multi-resistant staphylococci. But then vancomycin gained in significance as a result of the triumph of the toughened up staphylococci. In the meantime, in Europe, the vancomycin-resistant enterococci (VRE) had collected in people's intestines. In 1994, antibiotic expert Wolfgang Witte of the Robert Koch Institute in Germany found these bugs in 12 per cent of human faeces he analysed, as well as in numerous kinds of foods. Was it a ticking time bomb?

Thanks to the restraint of European doctors, VRE in hospitals was not – nor is it today – the huge problem it was in the US, where doctors had generously prescribed vancomycin. Yet the impressive number of VRE carriers in the European population threatened to change this. For this reason, avoparcin was banned in Denmark in 1995, in Germany in 1996 and throughout the EU in 1997. The number of human VRE carriers, which Witte investigated again in 1997, promptly dropped to 3 per cent.[34]

In 1999, the EU finally banned all growth promoters, with the exception of four reserve drugs that, as of 2005, are also illegal. Despite this, antibiotics have not completely left the barns. They may still be used as drugs, which means that fattening animals will also continue to be fed large amounts. In 1997 alone, 3400 tons of the microbe killers were administered to animals in the EU for treatment and prevention. Even today, critics are asking how well the new rules will eliminate abuse.

## Fatal evidence

The ban on fodder containing antibiotics in the EU was passed in the face of massive opposition from the agricultural industry. In the US, the industry has managed to successfully fight the imposition of a similar ban so far, arguing that no one has been able to produce direct evidence that a bug created through growth promotion has made a person sick. While it is almost impossible to provide this kind of proof, there have been some cases where researchers, with the fervour of detectives, have come close. 'The evidence justifying the ban has been around long enough', says Witte of the Robert Koch Institute.

The evidence was clear enough for the infected victims as well – and fatal. A case in Denmark attracted so much attention that it had a major impact on the EU's decision to issue the ban. Twenty-five people fell ill from *Salmonella typhimurium* DT104, a notorious diarrhoeal microbe that has often made the headlines. Two people died. DT104 has a set of resistances to five antibiotics. In 1996, this powerful variant was responsible for 96 per cent of all cases of salmonellosis (food poisoning) in Germany, and in England and Wales it accounted for 43 per cent of cases in 2000.[35]

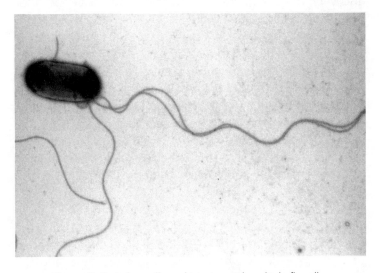

**Figure 2.3** *Salmonella typhimurium* with multiple flagella, as seen through an electron microscope

The DT104 microbes that hit Denmark in June 1998 were worse. Somewhere along the line they had picked up two more resistance genes. Although doctors had been warned about DT104, this expansion of the arsenal took them by surprise. For a 62-year-old woman who finally dragged herself to a Copenhagen hospital after suffering from severe diarrhoea for nine days, it was lethal. The doctors treated her with ciprofloxacin, which usually wipes out *S. typhimurium* quickly. It was completely ineffective. The *S. typhimurium* penetrated the intestinal wall in her body and inundated the organs. She died soon afterwards.

At the same time, researchers in the monitoring network established in Denmark to contain DT104 made an important discovery. In a slaughterhouse on the island of Seeland, they stumbled on an unusual variant of DT104, with seven resistances rather than five. It was the same bug found in the 62-year-old woman and four other patients. The researchers launched a dragnet operation. They phoned slaughterhouse workers, patients and butchers. By the evening, the puzzle had fallen into place. All the patients had bought their pork from butchers who had purchased their meat from the slaughterhouse in question. Shortly after, the biodetectives found the infected herds. While the pigs hadn't been treated with the antibiotic, some herds belonging to neighbouring farmers had.[36]

Apparently, the salmonella had jumped from one of these nearby farms to the herds that were later butchered. Also, the killer bugs wandered from infected people to other victims. In addition to patients who had been infected by eating the meat, one of the victims was a nurse. The microbes know no barriers. They jump from animals to humans, humans to animals, they get into water via animals, from water to the soil and vice versa. And everywhere they encounter antibiotics that encourage resistance. Cargo by the ton enters the environment via hospital sewage pipes or slurry sprayed onto fields by farmers. A study carried out by the Swiss Federal Institute of Aquatic Science and Technology in Dübendorf showed that every time slurry fertilization takes place, up to 500 gm of antibiotics per hectare of field are spread.[37]

## A single mutation, and $100 million go down the drain

Whether the bombardment with antibiotics comes from contaminated soil or water, the feed trough or an infusion is irrelevant. The constant threat steadily challenges the bugs to defend themselves. Ever since molecular biology has made it possible to view the inner workings of microbes, researchers have watched them as they do this and discovered how efficient and creative these survivors' defence mechanisms are. This can be demonstrated by using the example of penicillin and staph. In order to grow, the bacteria need to continuously build up their protective cell wall, which is woven out of a network of different molecules. This is where penicillin intervenes. It blocks an enzyme that connects a strut between newly inserted molecules. Because the struts in the membrane are suddenly gone, the bacterium bursts.

Other antibiotics exploit microbes' other weak spots. For instance, they hamper the essential production of proteins. However, the harassed microbes find an antidote for every poison. They destroy the penicillin with their own enzyme, beta-lactamase, or they use powerful pumps to transport other drugs from their interior.

Part of the bacterial defence arises by coincidence. Mutations are always occurring in the genetic material of each bacterium. Sometimes they modify genes, which contain the blueprints for enzymes and in turn change the corresponding enzymes. Sometimes this type of mutation hits an enzyme that is an antibiotic's target of attack. If the modification affects the enzyme's construction, such that it continues to be functional but can no longer be attacked by the antibiotic, the microbe will become resistant.

This mechanism saved many bacteria when the researchers threw newer and newer modified penicillins into the battle. After each new generation was introduced, microbes turned up whose beta-lactamase also destroyed the new foe by muta-tion. 'It is frightening to realize that one single base change in a gene encoding a bacterial beta-lactamase may render useless $100 million worth of pharmaceutical research effort', lamented Canadian antibiotics expert Julian Davies in *Science*. That's how much it can cost to develop a new drug. Davies' estimate was a conservative one, though, since these days the costs are often closer to $500 million.[38]

Still, the majority of the microbes' assortment of resistances was probably already there before humanity began swinging its chemical mace. In the short time in which antibiotics have been used, it is unlikely that entire enzymes could have emerged such as the extremely effective pumps that transport penetrating antibiotics out of the microbe again. Researchers assume that some resistance enzymes were originally assigned to other duties. For example, some pumps served to dispose of poisonous substances taken in from the environment. When humans began the chemical war, these enzymes suddenly developed another life-saving function – for the microbes.

Many resistance enzymes probably stem from the same source from which humans get most antibiotics: many antibi-otics are produced by micro-organisms, which probably use them to reserve a small place for themselves in the microbe jungle of the soil. In order not to be killed by their own poison, antibiotic producers have to protect themselves with resistance genes. For instance, in addition to genes for vancomycin production, the soil organism *Amycolatopsis orientalis* also hosts a whole group of resistance genes that show up again in VRE. It appears that the exchange of genetic material allows the protective genes to get from the antibiotic producer to the enterococci.[39]

## Threatening sex

The extent of bacterial promiscuity that bacteriologists are gradually discovering amazes even the most callous members of the guild. The microbes incorporate pieces of genetic material from other dead organisms that are lying around or get them from viruses or through sex. In the case of the bacterial type of mating, thin tubes take on the role of the penis. They are used to exchange genetic material. In rare cases, the microbes even fool around with more high-class beings like yeast or plants. Virginia Waters, of the University of California, San Diego, recently carried out the ultimate sex experiment. She offered hamster cells to bacteria. Even this partner wasn't spurned by the sex-obsessed microbes, and they conferred parts of their genetic material. All Waters had to do was to give them enough time – one night.[40]

In 1959, scientists first became aware of the horrific speed with which the bugs distribute their resistance genes. The incident took place in Japan, where a variant of the bacterial dysentery agent *Shigella dysenteriae* emerged that was immediately resistant to four antibiotics. In the same diarrhoea samples, scientists also found *Escherichia coli* cells that were resistant to the same substances. The researchers quickly realized that the two fourfold resistances could never be the result of independent mutations, since the probability of that happening was too small. There was another explanation: in

contrast to most genes of a bacterium, which sit on a chromosome, a single long thread of genetic material, these resistance
genes were piled on a small, circular piece of genetic material,
so-called 'plasmids'. This is why they could be efficiently transferred from one microbe to the next in a single go. The
microbes immediately showed just how efficiently they could
do this. Not long after the discovery, bugs carried so-called 'R-
factors', as scientists had dubbed these specific plasmids, to
the ends of the earth.

In addition to R-factors, microbes can host an amazing
selection of other plasmids. They are like attics, where the
bacteria save what they don't need every day but which can
be helpful in special situations: genes for enzymes that can
destroy exotic substances or resistance genes against heavy
metals. The devilish plasmids existed even before antibiotics
entered the medical scene, however. In the depths of their
freezers, researchers found a vial with bacteria that had
been deep frozen in 1946. The microbes had hoarded resistance genes against tetracycline and streptomycin on an
R-factor, although no doctor had ever used the substances.
However, the analysis carried out by a group of researchers
who examined bacteria collected from human faeces around
1930 showed that, while the bacteria possessed R-factors,
they carried very few resistance genes. The structure was in
place. All that was needed was a push to set off the
avalanche.[41]

One of the avalanches currently gathering momentum as a
result of plasmids makes Robert Koch Institute researcher Witte
nervous. In order to counterattack penicillin resistance, in the
1980s, pharmaceutical companies launched several new classes
of antibiotics on the market that defied the resistance-
conferring beta-lactamase of the bacteria. While some of the
more exotic bugs did a dry run at rebellion, it didn't seem to be

that dramatic. The genes responsible were stuck to the chromosome, which hindered exchange between the microbes.

Suddenly, in 1990, *Klebsiella* bacteria that hosted the mutated beta-lactamases on a plasmid turned up in Greece. The danger was that a resistance that knocked out several substance classes simultaneously could spread among a wide range of bacterial species. And that's exactly what happened. In 1991, the dangerous cargo appeared in other bacteria in Japan. This occurred again in France and Saudi Arabia in 1992, in Guatemala in 1993 and in the US in 1994. Today, they have become entrenched almost all over the world.[42]

How these broad-spectrum resistance genes managed to jump from the chromosomes to the plasmids remains a mystery. Yet bacteria also have their tricks for making such leaps. For instance, there are pieces of genetic material that can hop from one position in the chromosome to the next or into a plasmid. One of these so-called 'transposons' appears to attract resistance genes in particular. This enables entire colonies of resistance genes to emerge that present their owner microbes with multi-resistance.

## 'The pre-antibiotic era has returned'

The whole arsenal that bacteria have acquired and are constantly developing leaves no doubt that antibiotic resistance is not going to disappear. Microbes quickly counterattack even brand new substances, as shown by the almost immediate emergence of staph armed against linezolid. It is true that doctors can improve the situation by calculating the dosage of antibiotics more precisely. For example, in Scandinavia and Switzerland, the resistance situation is far more under control than it is in Southern Europe, where antibiotics are used much less specifically.

However, in some countries, where until recently the situation looked good, resistance is also on the rise. This trend is helped by holidaymakers who bring resistant staph home with them from Spain, immigrants from former Soviet Union countries that are experiencing TB epidemics or businesspeople who inadvertently bring multi-resistant cholera agents from Africa. The bacteria are globalization fans. An investigation carried out by a team headed by Alexander Tomasz of Rockefeller University in New York City revealed that 70 per cent of all multi-resistant staphylococci from hospitals worldwide belonged to only five strains, and they could all be traced back to only two lines of origin.[43] In UK hospitals, two strains of MRSA are primarily responsible for the epidemic raging there.[44]

In the past 20 years, pharmaceutical companies have greatly reduced research on antibiotics. At least in wealthy countries, infectious diseases appeared to be under control, and more money could be earned from drugs for cancer or 'diseases of civilization' such as obesity. These days, the consequences of this profit optimization are clear. New substances for stopping the flood of microbes are lacking. 'Unless things change soon, we are going to face a major health crisis', warned Brad Spellberg, an infectious disease researcher at Harbor-UCLA Medical Center in Los Angeles, in *Nature*. Even if pharmaceutical companies immediately cranked up their antibiotic research, he said, the long development period means that it will take years for more drugs to enter the market.[45]

In developing countries, many more people are noticing what happens when an infection no longer responds to any drug. In India, over two-thirds of the cases of typhoid fever are resistant to chloramphenicol, once the drug of choice. Now quinolones (for example ciprofloxacin) are being used, which are much more costly, putting them out of reach for many patients. Yet even these drugs have a 20 per cent failure rate. The treatment of multi-resistant tuberculosis, which is raging in Russia, Asia and Africa, costs 100 times more than the treatment of the normal variant. Every year, 3.5 million people die of a respiratory infection and over 2 million succumb to diarrhoeal diseases, both of which are often caused by bacteria. They are joined by 2 million deaths caused by tuberculosis.

However, in England and the US, there are also cases where no amount of money can help. People are dying because there are no effective drugs. The subtitle of an article about the wide spread of multi-resistant *Acinetobacter* and *Pseudomonas* bacteria in downtown New York City summed up the bitter realization: 'The pre-antibiotic era has returned.'[46]

What are needed now are new ideas. Researchers have found new active agents in human perspiration and the mucous membranes of frogs. Others are working on novel substances in chemical laboratories. And then there is an exotic treatment method that harks back to the beginning of the last century: phage therapy.

# 3

## the wild pioneer era

On the evening of 1 August 1919, an exhausted Robert K was admitted to the Hôpital des Enfants-Malades in Paris. The 11-year-old had already gone to the toilet 12 times that day, and 12 times the only thing that came out was liquid and bloody mucous. The doctors' diagnosis was bacillary dysentery, a serious illness caused by *Shigella* bacteria. Victims suffer from diarrhoea, high fever and bouts of severe abdominal pain. The toxins of the microbes can trigger a collapse of the blood vessel system and shock, among other symptoms. Back then, diarrhoea was often a death sentence.

But Robert was lucky. Two days before he was admitted, a researcher at the Pasteur Institute in Paris had been to see the head of the department, Professor Victor-Henri Hutinel, explaining that he had a new remedy for treating dysentery. His name was Félix d'Herelle, and he told Hutinel that he had discovered a previously unknown microbe, which, in a spectacular manner, dissolved even the densest *Shigella* cultures in a matter of hours and rapidly multiplied as they did so. During his investigations, the enigmatic microbe had always appeared in the stools of dysentery patients just before they began to recover from the disease. This is why d'Herelle said it had something to do with the healing process. He had called his discovery 'bacteriophage'. Now he wanted to test its healing properties in some of Hutinel's patients by giving them 2 cc of bacteriophage culture.

Félix d'Herelle later described these events in great detail in his memoirs, which to date remain unpublished:[1] Hutinel accepted the offer, but on condition that d'Herelle proved the remedy was harmless. D'Herelle told the paediatrician that he would drink a dose 100 times greater than the one he planned to give Robert K. 'I had already been taking large amounts of these types of solutions, and afterwards all my family members tried it as well. I was able to confirm that the bacteriophages would pass through the alimentary canal without causing even the slightest side effect.' The next morning he brought a full flask to the hospital. Twenty doctors, including Professor Hutinel, tried the 'honorary bacteriophage', as one of the assistants called the drink. The circle of academics declared that while it wasn't particularly tasty, it wasn't that bad, either. This type of safety test wasn't uncommon then. Robert K was admitted that evening.

The next morning, at 10 o'clock, d'Herelle gave Robert 2 cc of bacteriophages. The idea was that they would do what they had done countless times in the lab – dissolve the germs – but this time in the boy's intestine. And they appeared to do the job. During the afternoon the boy only had three bloody bowel movements. During the evening he moved his bowels once, and the stools were loose but not bloody. The next day the symptoms completely disappeared and the dysentery bacteria were no longer in the stools. D'Herelle and the doctors observed Robert for another week and discharged him as cured.

For nearly a month, no other dysentery patients were admitted, but then on 28 August four patients were admitted to the hospital with severe dysentery. Three of them were brothers whose sister had already died from the diarrhoea. These children were also treated with bacteriophages and were all cured.

In line with the experimental nature of the unusual treatment, in a later publication, d'Herelle warned that the small number of cases was not enough to provide 'absolute proof' for the method. However, he felt that his observations substantiated his expectations that the bacteriophages multiplied in patients' intestines at the expense of the bacteria: he detected them in patients' stools as long as they were sick and as long as he could detect the bacteria. But when the dysentery bugs disappeared, so did the bacteriophages.

Curing the five children began a new chapter in the annals of medicine. It was a tumultuous era in which doctors would soon be arguing about how well the bacteriophages got rid of infections. The loudest voice was that of the discoverer himself. No one had a bigger impact on this first period of phage therapy and its reputation as d'Herelle – both positive and negative. He was a charismatic person, with a burning passion and know-it-all arrogance, a thirst for adventure that took him to the limits and downright courage. D'Herelle travelled to the ends of the earth in the fight against plague and cholera and for years was involved in quarrels that are among the most bizarre in the history of science. He was a genius, but one who spent his life rubbing people up the wrong way.

The unusual history of phage therapy provides insights into the course of science, which is sometimes as affected by passions, dreams or envy as by rationality. The pioneers' struggle also contains many lessons that are relevant today, when it comes to resuming the use of bacteriophages to fight bacteria.

## Death everywhere

D'Herelle had begun the hunt for the mysterious microbes during the First World War. He had been working at the Pasteur Institute since 1911. Once war broke out, however, manufacturing vaccines for the Allied soldiers sidelined all other projects. Along with his colleague Alexandre Salimbeni, d'Herelle produced over 12 million doses of vaccine. At times his wife and two daughters helped with production. D'Herelle did research on his bacteriophages between 6 pm and 1 am.[2]

**Figure 3.1** Félix d'Herelle, his wife (left) and two assistants (right) at the Pasteur Institute during the First World War

The prospect of a cure for bacterial dysentery and perhaps even other epidemics was a strong motivation for d'Herelle's night shifts. Infectious diseases were still major killers, and very few drugs were effective: Edward Jenner's smallpox vaccination that he had discovered in the early 19th century, a few antiserums and vaccines along with Salvarsan for syphilis, which had been put on the market in 1913. However, Salvarsan, an arsenic compound, had serious side effects.

A look at *Merck's 1899 Manual* for physicians reveals the dearth of proper remedies available to doctors for treatment at the time: if you look up 'diphtheria', alongside Emil Behring's famous antiserum, 74 other 'healing substances' are listed. There are a few ineffective but at least harmless remedies like lemon juice listed next to toxic substances such as arsenic or mercury. Even strychnine, the murderers' poison of choice, is mentioned in *Merck's Manual* several times. One disease it was supposed to help treat was tuberculosis. Between the letters A and D, 133 diseases appear. Arsenic is recommended for 44, mercury for 42, and strychnine and cocaine for 20. To treat the clap (gonorrhoea), the doctor had 96 substances to choose from, all of which – as we now know – were ineffective.[3]

Prior to 1900, diphtheria was the most frequent cause of death for German children. In 1900, 'the strangling angel of children', along with scarlet fever and whooping cough, laid 65,000 children in the German empire to rest. The killer trio was the cause of 5 per cent of all deaths in Germany. Adults were also threatened by infectious diseases. In the same year, tuberculosis was the main cause of death in Germany, claiming 122,048 victims. Pneumonia came in third, with 76,497 deaths.[4] In the US, the order was the reverse: most people died of pneumonia (40,362). Tuberculosis was the second leading cause of death that year, killing 38,820. The case was similar in England, although exact figures are not available, because diseases were classified differently. Death statistics include antiquated terms such as 'miasmatic' or 'zymotic' diseases. Average life expectancy in 1900 was around 45 years. The First World War accentuated this picture. In England, for example, the mortality rate for tuberculosis rose dramatically, due to cramped living conditions and malnourishment brought about by the war.[5] Soldiers suffered even more. Thousands of wounded soldiers were threatened by gas gangrene and, in the muddy trenches, dysentery – the 'war epidemic par excellence' – was rampant.[6]

## Efficient killers

When d'Herelle announced his discovery in the middle of the First World War, at first it went unnoticed. Emile Roux, his superior at the Pasteur Institute, read the short paper at the meeting of the Académie des Sciences held on 3 September 1917. In the publication, 'On an invisible, antagonistic microbe of the dysentery bacillus', d'Herelle described his discovery, the microbe he had obtained from the stools of several patients recovering from dysentery.[7] Isolating it had been quite simple. A test tube full of broth, mixed with three or four drops of the stool sample, was incubated for 18 hours at 37 °C. The contents were then poured through a ceramic filter with holes so tiny that all the bacteria were held back. Just a few drops of the filtered solution could then completely destroy a culture of dysentery bacteria within hours or days.

The only thing d'Herelle could see of this microscopically tiny mass murder was that the murky soup turned into a clear liquid. Bacteriophages were much too small to be seen through any microscope, until 1939, when Helmut Ruska was the first person to spot them through the brand-new electron microscope.[8] He saw 'little round bodies' sitting on the outer wall of the bacteria. They were tiny, only about one ten-thousandth of a millimetre in size. Only later electron microscopes with higher resolution showed all their details: the prototype of a bacteriophage has a head, the 'little round body' that Ruska had seen, sitting on a tail that has more or less long tail fibres sticking to it at the other end.

**Figure 3.2** A T4 phage that infects *E. coli* cells,
as seen through an electron microscope

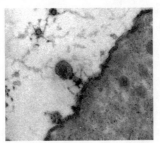

**Figure 3.3** A T4 phage on the cell membrane of an *E. coli* cell,
as seen through an electron microscope. The phage has already
introduced its DNA into the bacterium

In the past 50 years, in addition to electron microscopes,
modern methods of molecular biology were needed for virologists to thoroughly investigate bacteriophages. Phages are
viruses that only attack bacteria. They have only one objective:
their reproduction. They are so poorly equipped, however,
that they can't do this on their own. Without the help of their
victims, they aren't much more than a dead piece of protein
with a touch of genetic material. But if the phages do hit a
suitable bacterium, they multiply in a chillingly efficient cycle.

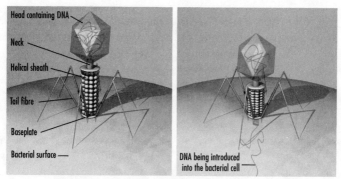

Head containing DNA

Neck

Helical sheath

Tail fibre

Baseplate

Bacterial surface

DNA being introduced
into the bacterial cell

**Figure 3.4** A phage docking on the surface (left)
and introducing its DNA into a bacterium (right)

The reproductive cycle begins when the virus uses its tail
fibres to attach itself to its victim. The details of what happens
next vary according to the different phage types. But their aim
is always the same: to get their genetic material, which is
located in the head, inside the bacterium. T4, a well-studied
phage infecting *E. coli*, then contracts its tail sheath which
pushes a tube located within the tail through the membrane of
the bacterial cell. The phage's DNA is passed through the tube
into the cell, where it takes control, brutally stops many of its
vital functions and forces it to churn out new virus components – heads, tails, tail fibres – in production-line style.
Then comes the final assembly. Finally, enzymes dissolve the
wall of the bacterium from the inside and the newborn bacteriophages reach the exterior, ready to attack new victims. The
viruses proceed very selectively as they do so. Most of them
attack only a subgroup of a single bacterial species. Generally,
they don't touch animal or human cells, which is why they are
harmless to human beings. (Some researchers say there are
extremely rare exceptions to this rule. This is explained in
further detail in Chapter 7.)

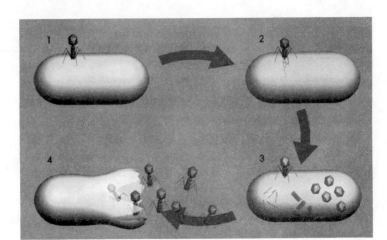

**Figure 3.5** Phage reproductive cycle: (1) A phage attaches to the surface of a bacterium. (2) Its sheath contracts, and its DNA is transferred into the bacterium. (3) The bacterium starts producing phage proteins and DNA. They are assembled into new phages. (4) Special phage enzymes dissolve the bacterial cell wall and new phages are released

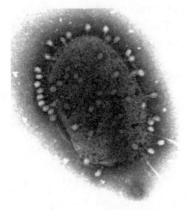

**Figure 3.6** Several T4 phages on the surface of an *E. coli* bacterium, as seen through an electron microscope

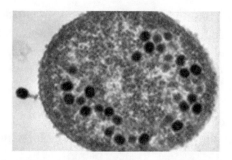

**Figure 3.7** Phage heads inside a bacterium. The phage on the outside was too late: once one phage has injected its DNA into a bacterium, the cell wall is altered such that other phages can no longer inject their DNA. The sheaths of these 'latecomer' phages contract, but the DNA remains inside their head

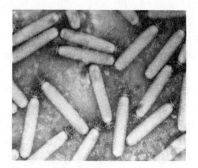

**Figure 3.8** Phages of the bacterial species *Lactococcus lactis*, as seen through an electron microscope. These bacteria are used in the production of dairy foods

Under ideal conditions, the reproductive cycle of the most virulent phages takes 30 minutes, during which time up to 200 viruses are produced per victim. In no time, billions of new viruses can be produced in a bacterial culture.[9] D'Herelle knew none of this when he made his discovery. He only knew that whatever was destroying the bacteria must be smaller than they were, because his ceramic filter held back the bacteria, but the organism that destroyed them passed easily through the holes.

These ceramic filters with tiny holes were developed by Frenchman Charles Chamberland in 1884 in order to remove the typhoid fever agent from contaminated water. The new device led to some revolutionary discoveries. In 1892, Russian scientist Dimitri Ivanovski described a filterable 'something' that caused mosaic disease in tobacco plants. In 1915, British bacteriologist Frederick Twort stumbled upon a filterable agent that dissolved bacteria. This means he made the same discovery as Félix d'Herelle, but two years earlier.[10] However, Twort's discovery fell by the wayside because of the turmoil of the First World War, only to reappear a few years later, leading to a furious controversy over who first discovered bacteriophages.

In d'Herelle's time, knowledge about viruses was minimal and the blueprint of the tiny microbes was unclear. This makes d'Herelle's degree of understanding that he demonstrated in the first short description of his experiments, contained in a mere two pages, even more amazing. He reported that the bacteriophage was a living microbe because it constantly reproduced itself. One drop of a dissolved bacterial culture was sufficient to decimate a new culture within a matter of hours. This could be repeated as desired. The phage from dysentery patients did not grow on dead bacteria or other types of bacteria.

The claim that his discovery represented a tiny microbe was bold, because he could have had a simple disinfectant substance in front of him. He offered evidence that a living organism was responsible for the effect by distributing a drop of a fresh mixture of a few phages and bacteria on a solid surface of nutritive gelatine. After 12 hours, d'Herelle saw a so-called 'lawn' of bacteria with several holes. He concluded that individual bacteriophages had landed on these places and reproduced at the expense of the bacteria. D'Herelle argued that a chemical substance could never concentrate at one place. With this, he had also discovered the fundamental method for isolating phages that is still used today – and had planted the seed that would generate years of bickering.

**Figure 3.9** Bacteria grow into a continuous covering on nutritive gelatine except at the spots where a phage was deposited

D'Herelle stressed that the appearance of viruses in faeces coincided with the recovery of dysentery patients. This led him to make another ambitious claim: his 'antagonistic microbe' triggered the cure of dysentery in patients. With this statement, as early as in his first publication, d'Herelle trained his sights on phage therapy, the great passion that drove him for the rest of his life.

## D'Herelle roots around in chicken manure and saves the feathered beasts

Once d'Herelle had observed the efficacy of phages in the test tube, it was time to try them out on animals. He wasn't daring enough to start testing humans and first wanted to confirm his results in an animal epidemic.[11] In spring 1919, a colleague at the Pasteur Institute told him about a poultry epidemic raging in the countryside. D'Herelle rushed to see for himself. He found a highly contagious epidemic that rendered the chicken coops a place of diarrhoea and death. A large number of the fowl had already died.

The bacteriologist took a few dead chickens back to his lab and discovered that the disease was not what the veterinary authorities thought it was. They had assumed that chicken cholera was decimating the flocks, but d'Herelle discovered *Salmonella gallinarum* bacteria in the chickens' blood, the agent of fowl typhoid, previously unknown in France. As was his wont, d'Herelle delighted in pointing this out in his publication. This habit, which he and his friends considered to be his passion for truth but which came across as arrogance and a know-it-all attitude to his enemies, would overshadow his life at various times and did not help to improve the reputation of his 'baby' – phage therapy.

D'Herelle next investigated the origin and distribution of the epidemic. It had already affected 14 French *départements*, was 'extremely lethal' and had a 'crushing' course. These aspects appealed to the microbe hunter, since the more spectacularly a disease raged, the more tempting it was to deal with. As he had done with dysentery, d'Herelle analysed the role of phages in the course of fowl typhoid in four chickens. To do this, he examined chickens' faeces for *S. gallinarum* and phages during the course of the disease. Again he concluded that the appearance of phages was required for the chicken to recover. When he followed the epidemic at other farmyards, he made an amazing discovery. The appearance of phages that attack *S. gallinarum* in a chicken's faeces not only heralded its recovery but also ended the epidemic at the particular farmyard. The other chickens pecked the virus from the manure and, as they did so, vaccinated themselves against the disease.

This was an extremely suggestive picture: phages, a healing viral epidemic, were counteracting salmonellosis, a lethal bacterial epidemic.

D'Herelle threw himself into experiments to test his daring hypothesis. In his lab, he first fed phages to a healthy chicken and then infected it with *S. gallinarum*. The chicken stayed healthy. Two chickens put in the same cage soon excreted phages in their faeces. After d'Herelle infected them with a fatal dose of *S. gallinarum*, the result was once again – nothing. Two chickens that had not been given phages died five days after being infected with the same dose of *S. gallinarum*.

In the *département* of Aube, d'Herelle now set up prevention tests at several farmyards. At about the same time, he launched his therapeutic trials at the Hôpital des Enfants-Malades. The agenda for the rest of his life had been set. From then on, filled with passion and commitment, he threw bacteriophages into the war on epidemics. For d'Herelle, who saw himself in the tradition of Louis Pasteur, the distance from the lab to the sickbed was no further than it had been for Pasteur himself.[12]

But first he was drawn to the hunt for new microbes. When Alexandre Yersin, discoverer of the plague bacterium and head of the Pasteur Institute in Saigon (present-day Ho Chi Minh City, Vietnam), came to Paris in 1919, there was no holding d'Herelle back: 'Indochina! That was the land of my dreams. Here you find cholera, plague and a whole range of animal epidemics like this horrible buffalo epidemic that in the past several years killed off all the buffalo on Java – more than a million – within a couple of months.'[13] On 6 March 1920, d'Herelle's daughter Huberte wrote in her diary: 'Papa has left for Saigon.'

## Some people do in one lifetime what others need five lives to accomplish

A life that began erratically continued in the same vein. When the 44-year-old researcher published his sensational discovery in 1917, the first turbulent half of his life was behind him, the events of which could easily have ruined two or three normal lives. Félix d'Herelle was born on 25 April 1873 in Montreal, Canada.[14] His father, 30 years older than his mother, died when Félix was six years old. His mother moved to France with Félix and his brother Daniel, who was five years younger than him. D'Herelle's grandfather had been born in France. Félix attended secondary school there. When he was 16, his mother gave him a bicycle and the proud sum of F1000 for a bike tour through the east of France, Germany, Belgium and Luxembourg.

During this trip, d'Herelle recalled later, something he experienced roused his interest in bacteriology. While breakfasting at a hotel, he heard that the day before a dog with rabies had bitten a boy who had been taken to a monastery in the nearby Belgian town of Saint Hubert for treatment. Félix wondered why the sick boy had not been sent to Paris to the legendary Louis Pasteur, who had an effective treatment for rabies. The hotel guests told him about the monks in Saint Hubert, who for centuries had successfully used relics to treat rabies patients. Félix jumped on his bike and rode the 60 km to have a look for himself.

This action characterized d'Herelle's entire life. He was always brimming with an insatiable curiosity about an endless number of topics. Half his memoirs are packed with descriptions of travels and studies undertaken on top of his actual research: investigations of history and customs, plants and animals. Everything interested him – and he had a distinct opinion about everything, too.

Despite this formative experience, in the beginning, Félix kept his distance from bacteriology. A second gift of F3000 from his mother allowed him to travel to South America for three months after graduation from school, then an unusual undertaking for a 17-year-old. The first stop on the trip was Tenerife. 'It was my first contact with the exotic. It captivated me', d'Herelle wrote in his memoirs, which he fittingly called 'The Travels of a Bacteriologist'.

On the way back, yellow fever broke out on the ship, but unlike the other passengers, who were in a state of panic, Félix was unflappable: 'I've probably had the most important characteristic of a good microbe hunter since the day I was born. I remain completely calm.' The tour through Argentina, Brazil and Paraguay was followed by travels in Europe. On a trip to Turkey, Félix met Marie Caire, the daughter of the French ambassador there. She became 'his wife and travelling companion'.

He continued travelling around Europe with her for a while until, at the age of 24, now the father of a daughter, he decided 'it was time to do something'. The family moved to Canada, where d'Herelle set up a home lab and read biological journals, as well as works by the British philosopher Sir Francis Bacon in the original Latin. During his solitary study of Bacon, he learned the scholar's scientific maxims that he would hurl at his numerous critics: 'You will not master nature unless you obey it.' Later d'Herelle would spell out the meaning of this phrase for microbiology countless times: scientists can only make valid conclusions if they study 'original diseases' – those who artificially infect lab animals with human agents will get worthless results. This meant that his study of chickens with naturally occurring fowl typhoid was sound, but a study involving unnaturally infecting rabbits, for instance, was not.

With this rationale, the autodidactic scientist attacked a good number of his research colleagues head on, because the so-called animal models of human diseases were and continue to be an important research method. With his demand that epidemics like cholera or plague be investigated in their natural surroundings, in exotic countries such as India or Indochina, d'Herelle combined his penchant for adventure with his scientific creed. Yet in a picture taken during this period, d'Herelle looks into the camera with such shyness that neither the intrepid adventurer nor the relentless critic can be detected.

In his home lab he taught himself bacteriological techniques. A friend of his father's arranged a job for him with the Canadian government. He investigated the fermentation and distillation of maple syrup to liquor. In his usual way of trusting only himself, d'Herelle didn't want to use purchased yeast for the experiments. Instead, in a long drawn-out process, he isolated his own yeast. In spring 1900, still lacking an actual profession, he escorted a group of geologists as a paramedic on a strenuous expedition to the icy Canadian northwest. At the same time, he and his brother invested in a chocolate factory, which soon went bankrupt. D'Herelle lost all but $2000 of his inheritance.

In the meantime, his wife gave birth to their second daughter, and it was now time for d'Herelle finally to start earning money. He applied to the government of Guatemala, which was looking for a bacteriologist, and he was hired – as the only applicant.[15] Guatemala was a country at the edge of the world then, the wild south. D'Herelle and a French chemist had the only research positions with the government. D'Herelle was appointed head of the bacteriology lab at Guatemala city hospital. After a short time, he was also assigned the task of developing a process for fermenting and distilling whisky from bananas.

For a Canadian family with two small children, life in a developing country was completely foreign. The d'Herelles had to

**Figure 3.10** Félix d'Herelle some time between 1900 and 1910

deal with the tropical climate, poisonous snakes and armed bandits in primitive surroundings. On arrival, the English ambassador recommended that d'Herelle carry a revolver with him at all times. It was a tip that paid off, when an escaped convict attacked him with a knife while he was riding a horse in a lonely area. D'Herelle emptied his gun into the thug's heart and kept riding because he didn't want anything to do with the dubious authorities.

At the end of the family's six-year stay, a yellow fever epidemic broke out in Guatemala. D'Herelle was appointed to work as a temporary doctor. It may be that the sight of people dying in agony made him all the more convinced that becoming a bacteriologist was a good choice. Yet at the same time he showed how stubbornly he could cling to his opinion. He had the houses of two families burned to the ground because they refused to comply with his hygiene orders.

Because of the yellow fever quarantine, the d'Herelle family suffered from acute starvation. Shortly before, they had all contracted malaria. Yet d'Herelle, ever the adventurer, looked back fondly on this time: 'When I think of Guatemala, a feeling of affection always comes over me. In this country I attended the school of hard knocks, and it was the beginning of my scientific career', he wrote as an old man.

In 1907 he received an offer from the Mexican government to continue his fermentation studies. This took him on a similar path to his hero, Louis Pasteur, who had also studied the fermentation of wine, beer and vinegar. The d'Herelles moved to the Chochoh sisal plantation near Merida on the hot Yucatan Peninsula. There d'Herelle investigated the fermentation and distillation of sisal, which is normally used for obtaining fibres. But sisal fibres were no longer in demand, and the plantation owners were looking for new applications for their plants.

The family suffered from the incredible heat. They were tormented by constant diarrhoea and frequent disruption of the water supply. On one occasion, they were all sick yet again. Félix kept throwing up, and the children had fevers and were skin and bone. On 10 July 1908, Marie d'Herelle jotted down a thought she had recorded so many times before: 'We are all very depressed.' Just before the family left Mexico, both girls contracted the dreaded yellow fever, although both survived it. Yet d'Herelle's work on the Chochoh plantation seemed to have been a success: he had managed to produce a tasty sisal liquor and had designed the distillery where it was to be produced in the future. The machines were ordered from Paris, where d'Herelle and his family went in spring 1909 to oversee their production. Since he had some time on his hands, he contacted the renowned Pasteur Institute and worked on an unpaid basis in the lab there in his free time.

Once the factory had been constructed in Yucatan, plans called for d'Herelle to become the director. The prospect of the unstimulating job prompted him to resign, however. Before he left the hacienda, his thirst for activity was quenched by a plague of locusts – a problem that was right up his street. He immediately thought of fighting the locusts with one of their own diseases. The head of the plantation helped him to pick through the locusts to find some that had diarrhoea, and he isolated a bacterium from their intestines. While he was still in Mexico, he carried out experiments to find out whether it would be suitable for stamping out the plague of insects.

After the d'Herelles finally moved to Paris in spring 1911, d'Herelle again felt the urge to work as an unpaid assistant at the Pasteur Institute.[16] On 22 May, the institute's director Roux presented the paper in which d'Herelle described his *Coccobacillus*, which had decimated the locusts, to the Académie des Sciences.[17] The lecture caused quite a stir, and the press reported on the ray of hope in the fight against the biblical plague. For the first time, the self-taught d'Herelle appeared in journals as the great white hope.

Not long after, an offer came from Argentina to test the method in the field. As a result, in late 1911, d'Herelle was once again far away from home. As an agrarian country, Argentina suffered miserably from the locust plague. After careful pilot tests, in which he tested the harmfulness of his bugs for other inhabitants of the pampa such as sheep and rabbits, in 1912 and 1913 d'Herelle led two campaigns against the voracious insects.

In the annals of the Pasteur Institute, he described his actions as very successful.[18] In Argentina, however, the results were controversial. Critics declared that the effect was not as good as d'Herelle described it. In his memoirs, d'Herelle angrily blamed this on his enemies in the Ministry of Agriculture,[19] and it even led him to quit. In other places, however, d'Herelle the insect exterminator was well respected. For instance, the Pasteur Institute in Algiers invited him to a campaign, as did the Pasteur Institute in Tunis during the war in 1915. Yet his method remained controversial. Some of its users saw successes, while others did not. British locust authority Sir Boris Uvarov criticized the method in a book published in 1928. Uvarov claimed that d'Herelle had been much too hasty in declaring it effective. He did, however, give d'Herelle credit for making biological pest control popular as a result of his experiments.[20]

D'Herelle reacted to the criticism with indignation, his usual manner. He saw himself as an outsider who had violated the dogmas of the establishment:

> Theories have no impact on me whatsoever. I observe and experiment. If my results coincide with theories, that's fantastic. Then I'll accept them. If they don't, I'll discard them, no matter what authority defends them. When I put forward a theory [myself], then I insist that it takes all the facts into account, that it explains them all and that it does not contradict any of them ... If it does, then it doesn't bother me, even if it is quite strange, i.e. contradicting the official theories. That's what has caused me the enmity of the 'official scholars' in all countries.[21]

At this point, he threw down the gauntlet to any insect researchers who viewed his campaigns critically. But before long, he would train his sights on other colleagues.

## The first disciples

The position of the busy outsider, who had started as an unpaid assistant, was by no means secure at the highly esteemed Pasteur Institute. When d'Herelle returned to Paris in late 1920, after his first therapeutic trials and subsequent journey to Indochina, his lab had been assigned to someone else. He was only able to continue his work because a colleague felt sorry for him. Biologist Edouard Pozerski donated d'Herelle a wobbly stool in his lab. It was a precarious situation for d'Herelle, who at 48 had to support his family and was still struggling for recognition as a researcher. In his memoirs, d'Herelle blames Albert Calmette, whose name is immortalized by the 'C' in the famous BCG (bacille Calmette-Guérin) tuberculosis vaccine. Calmette was the deputy director of the institute and in charge of administration. D'Herelle had made negative comments about the BCG vaccine because he suspected it was not entirely harmless. According to d'Herelle, Calmette hadn't forgiven him for this and had taken the lab away from him.[22]

Yet d'Herelle's fate and that of phage therapy soon took a positive turn. In 1921, despite his spats with Calmette, he was able to publish his first book, *Le Bactériophage – son rôle dans l'immunité*, in the series 'Monographies de l'Institut Pasteur', by quickly taking advantage of his mighty opponent's absence from the lab. In this first, larger publication, he described his theory of the essence and effect of phages and his studies on dysentery and fowl typhoid in great detail.

The new research area increasingly aroused the curiosity of other researchers as well. After d'Herelle submitted his report describing the use of phages in treating 11-year-old Robert K, Belgian researchers R. Bruynoghe and J. Maisin of the University of Leiden's institute for bacteriology went one better. 'We had the opportunity to use the staphylococcal bacteriophages in therapy and several beneficial results moved us to publish them', they dryly reported to the scholars of the Société Belge de Biologie on 3 December 1921.[23]

Approximately two years after d'Herelle's first therapeutic trials, the Belgian researchers injected six patients who were suffering from boils or the nastier carbuncles with 1.5–2 cc of staphylococcal phages in the area surrounding the affected places. In 24–48 hours, the boils emptied out and dried up. The authors concluded their presentation with a warning that the small number of patients treated was obviously not enough to evaluate the value of the method. And in the same dry tone they had used to present the report, the two researchers appealed to the imagination of their fellow physicians: 'We have tried this remedy in patients with carbuncles or furuncles, yet it is not impossible for the method to also be useful in treating the complications that arise from various skin diseases that are triggered by staphylococci.'

A number of groups rose to meet the challenge. On 28 January 1922, André Gratia of the Pasteur Institute in Brussels reported to the Société Belge de Biologie that he and D. Jaumain were working on a treatment for furuncles and carbuncles that involved staphylococcal phages. The two researchers had tested the therapy's effectiveness on rabbits before they tried it out on their patients. They also reported success, since the therapy brought about recovery that was clearly more rapid.[24]

Hopeful researchers quickly expanded the combat zone. Also in 1922, Paul Hauduroy of the University of Strasbourg's institute of hygiene and his fellow researcher A. Beckerich published their experiments on the dreaded typhoid fever.[25] They carried out several treatments in hospitals in Strasbourg and Orléans. This time the results were only mediocre. Of five patients who received the treatment, three were cured, but two died of the disease. The researchers assumed that this was because the doses were too small or the onset of treatment was too late. Early on, d'Herelle had repeatedly pointed out that phages could only save a patient if the toxins released by some bacteria have not badly damaged the victim's organs.

Research involving the healing viruses was not limited to d'Herelle's francophone home territory, but extended to other European countries as well. Richard Otto and H. Munter of the Robert Koch Institute for Infectious Diseases in Berlin had already investigated the therapy quite early on. On 29 December 1921, barely two-and-a-half years after d'Herelle's first experiments involving humans, they published an article entitled 'On the d'Herelle phenomenon' in the *Deutsche Medizinische Wochenschrift*.[26] It was mainly devoted to basic questions on the essence of phages. In two easily overlooked paragraphs, the two researchers reported on their own initial experiments with the therapy. Munter and Otto also first tested the curative power of their dysentery phages in animal experiments. They injected the viruses, along with a huge load of dysentery bacteria, into the abdominal cavity of a guinea pig, which survived the procedure. An unluckier fellow test animal succumbed after the researchers injected the bacteria without the phages. The animal experiments were a success.

Otto and his assistant Munter were not able to report the same outcome for a treatment they and a fellow colleague carried out on people. Professor Friedemann administered phages that were highly effective against dysentery and typhoid fever to several patients. Otto and Munter described it with Prussian brevity: 'With this method of treatment, the physicians responsible for treating the patients could not be convinced of an impact that corresponded to the expectations.' Otto decided that the reason for the setback was the caustic stomach acid, since other researchers had previously observed that phages are destroyed by acid. 'We have thus decided to test other methods of application.'

In 1925, Austrian Erich Zdansky tested the limits of imagination when it came to finding out the therapeutic powers of the viruses.[27] Zdansky was an assistant at the University of Vienna's

Medical Hospital Number 1. His research focus was the stubborn urinary tract infections caused by the intestinal bacterium *Escherichia coli*. This illness caused huge numbers of chronic cases that doctors could not control. 'The experimental material', as Zdansky called his patients, included 20 'cases', 15 of whom were in a chronic stage and had been pestered with a wide range of methods, all with no improvement. The energetic physician seemed to have searched out his difficult cases from all parts of Vienna. One wonders whether the subjects had actually been asked for their consent in participating in the experimental treatment. Zdansky only provided the following information in a footnote: 'Some of the patients were kindly supplied to us by Primarius Dr. R. Bachrach ... Privy Councillor Prof. Dr. Peham and Primarius Dr. F. Passini generously supplied us with one patient each.'

Zdansky's draconian regimen began with patients first being forced to drink lye to neutralize the acid in their urine that was harmful to phages. Afterwards, he used a catheter to rinse the bladder out with a saline solution before he pumped in up to two decilitres of phage culture. The patients were then instructed to retain the liquid in their bladders as long as possible. Zdansky repeated the procedure several times at intervals of one to two days. During this period, the patients' drinking was restricted in order to minimize the dilution of the phage broth in the bladder.

Using this complicated technique, Zdansky was able to cure six patients, which they truly deserved after being subjected to this unpleasant procedure. With the remaining 14 patients, *E. coli* germs that were resistant to the phages accumulated in the bladder. D'Herelle had already observed this phenomenon. When he grew dysentery bacteria in several flasks and then added phages, at first the culture became clear in all the flasks. However, in several flasks, the liquid became murky again because some germs proved to be resistant and continued to grow.

The resistant germs can be duped by setting an entire cocktail of different phages loose on them. Even if a bacterium is resistant to a certain virus, it can be overpowered by a different one from the mixture. D'Herelle always stressed the importance of these cocktails. There was another reason for this. He quickly realized that a certain dysentery phage may not destroy all the bacteria of the *Shigella dysenteriae* species, but only a few subgroups, so-called 'strains'. Each bacterial species consists of various strains that are more or less distinct from each other. In fact, when d'Herelle was doing his experiments, several bacteria were viewed as their own species that today are considered to be strains of the species *S. dysenteriae*. The differences between some strains and species can be so blurred that these days many bacteriologists are no longer willing to use the term 'species'. At any rate, d'Herelle found that mixing dysentery phages with different 'appetites' cured a larger number of cases of dysentery.

These early studies have already shed light on the problems of phage therapy that need to be solved if biological pest control is to work in the human body. One of the many issues is Otto's observation that viruses do not survive a bath in stomach acid. This is why, today, phage therapists in Georgia give their patients sodium bicarbonate, which neutralizes the stomach acid.

Zdansky avoided another problem by pumping the phages directly into the bladder to ensure they arrived at the right place. Phages may be tiny, but in comparison to chemical substances they are quite large. You never know which nooks and crannies of the body they will penetrate and which they won't. Carl Merril of the US National Institutes of Health (NIH) says: 'Even today there aren't enough studies addressing this issue.' Zdansky's idea of manoeuvring the viruses to the site of application wasn't at all bad. In the case of Alfred Gertler's trapped microbes, the Georgian doctor also used IV tubes to flush the phages as closely as possible to the bacteria in his foot.

However, Zdansky reached his limit when it came to the resistant bugs. This is still a problem today, as is the phages' extremely picky appetite. Some researchers follow d'Herelle's lead and continue to use virus cocktails, but they are not in the good books of Western pharmaceutical approval authorities such as the FDA. This has caused other scientists to pursue an alliance between the old phage technology and the new area of genetic engineering.

## Triumph

After the first tentative experiments, phage therapy spread like wildfire. Doctors in Italy, Spain, Holland, Denmark, Sweden and the US began using it. For d'Herelle, it must have been an exciting time. As an outsider, and practically single-handedly, he had created a new area of medicine.[28] The number of research papers rose. While the early 1920s saw some 20 per year, in 1930 there were almost 60 papers on phage medicine. Scientists groaned at the flood of publications, which made it impossible to keep track of the booming field. Considering today's avalanche of information, the complaints are almost touching.

In France, the home of phage therapy, physicians made brisk use of the new remedy. In the US, it also had quite a fan club. In 1930, Thurman Rice, a doctor at Indiana University in Indianapolis, described 300 cases.[29] These patients were plagued by all sorts of purulent inflammations, most caused by staph. In 90 per cent of cases, Rice was able to report success – an impressive result.

He was even able to cure 10 children with 'generalized furunculosis of extreme grade'. The task that the phages had to carry out to pull this off sounds herculean. The patients were between five months and ten years old and were covered with up to 350 boils. At the onset of treatment, most of them were in a pitiful condition. They were emaciated, malnourished and plagued by high fever. Some of them were on the verge of death. Rice applied the phages in wet bandages or injected them directly into the affected areas, which must have been a painful procedure. Rice saw a 'spectacular improvement' in all the children shortly after the onset of therapy.

However, phage therapy's first big breakthrough took place in Brazil. José da Costa Cruz, of the Oswaldo Cruz Institute in Rio de Janeiro, began doing experiments there around 1920. His first experiments were a flop. D'Herelle stepped in and gave the sceptical scientist some advice. D'Herelle felt that the activity of da Costa Cruz's phages was too weak. He advised him to look for more potent phages. In summer 1924, da Costa Cruz published the initial results of his research efforts: after successfully treating 24 patients, he and his institute dramatically increased phage production.[30] Within a year, the Brazilian scientists were producing 10,000 vials, a huge number, which were sent all over the country. According to da Costa Cruz, from then on physicians in the states of Para, Pernambuco, Rio, Parana and Bahia routinely used phage therapy. In his later writings, da Costa Cruz proudly stated that urgent requests for more phages had been submitted several times. Doctors were particularly enthusiastic about the speed with which the remedy worked. They only needed to empty the vials into the mouths of tormented patients and, in a matter of four to eight hours, their diarrhoea had already been reduced.

Da Costa Cruz was only told about two cases that had ended in failure. However, it appears that he didn't make a point of insisting that every unsuccessful case be reported to him, which of course was essential for properly assessing the treatment. Da Costa Cruz euphorically expounded: 'The dysentery phage is by far the best therapy for dysentery known to date ... For this reason, we are absolutely convinced that we have saved the lives of a huge number of patients.'

## Drunk on success

These kinds of triumphs were greedily received by the public, plagued by experiences of infection. People were prepared to expect microbiologists to perform miracles. In previous decades, legendary microbiologists Louis Pasteur and Robert Koch had rung in the golden era of bacteriology, with their discoveries and several therapeutic successes such as Pasteur's rabies vaccine. Soon the newspapers began reporting the new treatment and phage therapy became a media star. On 27 September 1925, the *New York Times* wrote about bacteriophages in an article entitled 'Tiny and deadly bacillus has enemies still smaller'. The subtitle put it suggestively and to the point: 'Life or death may depend on it.' It was not the only report about d'Herelle and the phage miracle to appear in the *New York Times*.

In the June 1931 issue of *Ladies Home Journal*, well-known medical journalist Paul de Kruif described the therapeutic trials carried out by Gratia and Jaumain nearly 10 years before:

Gratia was shooting shots of staphylococcus bacteriophage under the skin of a woman dreadfully sick with a huge carbuncle. Tense hours of worry for Gratia and his co-worker, Doctor Jaumain, as the woman got much sicker; then suddenly the next morning a tremendous discharge of corruption from her wound, with new strength flowing through her. Three days, all better. Magical![31]

When it came to heralding the news about promises of a cure, the impact of the media was as powerful then as it is today. When Karl Wilhelm Röntgen discovered X-rays in November 1895, it took less than two months for the Viennese newspaper *Die Presse* to publish the first report on the discovery. The news shot around the world via the media and fired the people's imagination. Soon there were reports of supposed X-ray-proof underwear to protect the wearer's privacy from prying, high-tech eyes. There was a similar reaction when Robert Koch was reported to have found a remedy for tuberculosis, which was raging at the time. Koch was lauded in the press and received countless letters requesting his miracle drug tuberculin – which soon proved to be ineffective.[32]

Shrewd researchers like Pasteur ingeniously exploited people's hopes. When he was crafting a vaccine against anthrax, for instance, he staged a public demonstration on a farm in Pouilly-le-Fort near Melun on 2 June 1881. Dying sheep, with black blood running out of their mouths, were put on display. They had been infected with the anthrax bacillus. Next to them, sheep that had been saved by Pasteur's vaccine were grazing peacefully. Pasteur or his assistants had ensured that newspaper reporters, including a reporter for the *London Times*, would be present at the spectacle.[33]

D'Herelle's enthusiastic rhetoric also attracted attention. For instance, at the association of French surgeons, he announced that the bacteriophages would 'revolutionize bacteriology, the pathology of infectious diseases, hygiene and therapeutics'.[34] The name d'Herelle had chosen for his discovery turned out to be an advertising boon. The artificially constructed word 'bacteriophage' gained in the translation. In the English-language press, it was often referred to as the 'germ eater'. The name evoked an image of an invincible troop of midgets devouring its way through a pile of staphylococci trapped by pus.

In his memoirs, d'Herelle recalls the successful christening. It happened on the day the hypothesis about the effect of phages in curing dysentery came to him:

> In the evening, under the light, I was telling my loved ones about what I had seen: the dysentery bacilli devoured by a 'microbe of microbes'. My wife asked me, 'What are you going to call them?' And the four of us put our heads together. Name after name was suggested and then discarded again. Finally, after all the discussion we came up with the word 'bacteriophage', a word formed from 'bacterium' and 'phagein', the Greek word for 'eat'. It was the 18th of October [1916]. I remember it because it was the evening before my younger daughter's [Huberte's] birthday.'

In his first book about phages, however, published in 1921, he wrote that 'phage' doesn't mean 'eat' in the strict sense, but rather 'develop at the expense of something else'.[35] It was necessary for him to specify this to his fellow researchers at the time. The meagre amount of facts meant that no one was able to say exactly how the viruses destroyed their victims.

The novel *Arrowsmith*, written by American author Sinclair Lewis and first published in 1925, shows the position of phage therapy in the public eye and in science at the time. Lewis had already become famous after publication of his socio-critical novel *Babbitt*. Now he wanted to write a satire about the US healthcare system and doctors who fleeced their patients. He invented the figure of Dr Martin Arrowsmith, who climbs the ladder from country doctor to famous researcher. Because Lewis wanted the book to accurately reflect the latest events in medical research, he turned to Paul de Kruif for advice. De Kruif had done research at the renowned Rockefeller Institute for Medical Research in New York, but was about to switch to the field of scientific journalism. In 1926 he became well known after publishing his book *Microbe Hunters*, in which he vividly described the life and research of famous bacteriologists such as Pasteur and Koch. Lewis asked de Kruif to brief him on the latest trends. Thus, *Arrowsmith* features the McGurk Institute, a copy of the Rockefeller Institute, as well as all kinds of ultramodern lab equipment and protagonist Arrowsmith's hot research topic – phage therapy.

A number of d'Herelle's scientist colleagues also predicted a great future for the new method, demonstrated by the honours soon bestowed on him. In addition to an honorary doctorate at the University of Leiden, in 1925 the Dutch Royal Academy of Sciences in Amsterdam awarded him the Leeuwenhoek Medal, which is only given every 10 years. D'Herelle was especially proud of this award because Louis Pasteur, his inspiration and model, had received the medal in 1895.[36] Since the Nobel Committee operates with absolute secrecy, d'Herelle was probably not aware that he had already been nominated for the Nobel Prize three times. In 1926, no fewer than eight fellow scientists nominated d'Herelle for the most esteemed of all scientific honours.[37]

## Plague!

In *Arrowsmith*, Lewis selected a plague epidemic on a Caribbean island as the spectacular premiere of phage therapy and this struck a chord with the readers. Although the plague only occurred sporadically in the US at the time and no longer played a major role in developed countries, it stimulated people's imagination. And in countries like India, it continued to claim thousands of victims. The Black Death remained as fascinating for microbe hunters as it always had been. In 1920, when d'Herelle was in Indochina, he had hunted plague sufferers and phages. However, he only managed to isolate plague phages from rats.

Five years later he was luckier. In 1925, d'Herelle was working for the Conseil Sanitaire, Maritime et Quarantenaire d'Egypte in Alexandria. After his dispute with Calmette and getting the boot from the Pasteur Institute and a mere guest performance at the University of Leiden, d'Herelle finally had a prestigious position. The Conseil was directly administered by the League of Nations and was supposed to prevent cholera and plague from spreading from Asia to Europe. The organization was concentrating on the yearly pilgrimage of Muslims to the holy sites in Saudi Arabia because the experts were afraid that epidemics could be carried along with the flood of people. In a quarantine station in El-Tor on the south flank of the Sinai Peninsula, pilgrims returning from Mecca and Medina were detained for four days. In addition, the Conseil checked the ships passing through the Suez Canal.

In July, 1925, d'Herelle wrote to his daughter Huberte[38] that he had come across four cases of plague in which he could finally use his phages from Indochina. Once again, he was his own guinea pig when it came to doing safety tests, injecting 1 cc of plague phages under his skin. When no harmful reaction occurred, it was time to try it out on the patients:

Théodore Cas ... deck hand, 16 years old. On 12 July, upset stomach with fever: on the same day, isolated in the hospital. His condition rapidly worsened; on the morning of the 15th had an irregular pulse of 126, temperature 39.4 degrees, collapsed. During the night a lymph node swelled up under the right side of his lower jaw until it was the size of a hazelnut, sensitive to touch. Culture tests and the infection of a guinea pig indicated *B. pestis* [today *Yersinia pestis*, the plague agent].

On the 15th at 3:00 pm, I injected a cubic centimetre of phage culture directly into the lump. On the morning of the 16th, all symptoms had vanished apart from the lump. The patient was in good spirits; when I came round for a visit, he was sitting up in bed. His temperature was 37.2 degrees, pulse 70 ... I had ordered the nurses to monitor him carefully during the night. They assured me that they had not seen any reactions, no sweat, no agitation. Several hours after the injection, the patient stated that he felt much better, fell asleep and woke up in the morning with the claim that he had been cured.

A new sample from the lump showed that the bacteria had disappeared.[39] Victory! Although news of the success was only based on four experiments, they attracted worldwide attention and encouraged other researchers to tackle the plague with phages.

## A momentous recommendation

D'Herelle's publication started more than that. He gave a copy of it to A. Morison, the English representative in the Conseil.[40] He was enthusiastic about the findings and sent a passionate letter to C. Heathcote-Smith, the British consul general in Alexandria. In the letter, he described d'Herelle's successes, ending with the plea:

I see every reason to hope for a favourable result by this method of treatment even in pneumonic plague. If so, then the dread of plague is conjured. Anti-plague serum is useless as preventative ... The only true prophylaxis is deratisation, and the only true treatment is bacteriophage. D'Herelle has supplied already the Sanitary Administration with the necessary bacteriophage. I think India ought to arm itself. Also all other countries where plague prevails. All honour to d'Herelle.

Morison's enthusiasm launched a field trial of gargantuan proportions: the 'Bacteriophage Enquiry' in the British Indies, which lasted several years. It was a dream come true for d'Herelle. At first, however, the pioneer only sent his plague phages to the Haffkine Institute in Bombay. When researchers carried out tests with them, they didn't see any effect. D'Herelle immediately took unpaid leave and travelled to Bombay at his own expense, where he straightened out technical problems and isolated a new plague phage. Yet in animal experiments, the phage couldn't cope with the plague bacillus. D'Herelle's explanation for this was that the plague bacillus in India was 'extremely virulent'. He returned to his post in Egypt in a matter of weeks.[41]

Despite this failure, the microbe hunter and his therapy made a lasting impression on Lieutenant Colonel John Morison, the director of the Haffkine Institute. He urged the Indian government to invite d'Herelle to return, this time to find a solution for cholera. And the next spring d'Herelle rushed back to India. Like Indochina in earlier years, India – where plague, cholera and malaria were raging – was now the land of his dreams.

D'Herelle considered cholera to be the queen of diseases: 'Finally cholera', he wrote in his memoirs. 'I'm familiar with several "impressive" diseases, including smallpox and yellow fever, but cholera surpasses them all.'[42] The course of the disease is impressive. It begins with a phase of stomach cramps and dizziness. Left untreated, this is followed by violent bouts of diarrhoea, 20–30 times a day, and vomiting until there is nothing left in the intestines but liquid and pieces of its own wall. Then muscle cramps set in. Extreme dehydration brings about an unquenchable thirst. The eyeballs collapse. Pulse, blood pressure and temperature drop. At the time, the death rate was 60 per cent. India was the world's cholera hot spot. In the 1930s, it claimed the lives of 200,000 people annually. From there, epidemics spread worldwide. In the 19th century, it had also hit Europe, shocked and defenceless, four times.[43]

## Blitzkrieg in the land of epidemics

At this point, d'Herelle had expanded his theory of the healing function of bacteriophages.[44] In his view, phages were almost the only substance that could save the body from death by bacilli. As far as he was concerned, the immune system had nothing to do with curing a new infection. However, he did concede that the body's own immune defence was responsible for immunity acquired after recovering from an infection. This contradicted the accepted medical opinion of the period, which of course delighted him.

His daring theory was based on observations of dysentery patients in whom effective phages had appeared in their stools prior to recuperating from the disease. If the viruses didn't appear, the patients died. As in the case of fowl typhoid, he assumed, the healing power of phages was just as infectious as the disease itself. A fatal epidemic is pitted against a healing epidemic.

This philosophy was the foundation of d'Herelle's battle against cholera. The experiments he carried out in the short period from April to October 1927 with Major Reginald Malone, the assistant director of the Haffkine Institute, and the Indian physician M. L. Lahiri were designed according to his familiar recipe.[45] First, the three researchers observed 27 patients in the Campbell Hospital in Calcutta for the occurrence of cholera phages and the course of the disease.

D'Herelle was able to confirm his theory. In 7 patients, no bacteriophages were detected, or only phages with very weak activity, and all 7 patients died. In the other 20, a phage appeared before recovery. In order not to impact the results by their potential bias, the researchers divided up the tasks. Lahiri observed the patients and collected stool samples, Malone identified the cholera agents in the samples and d'Herelle did the phage experiments. This means that Malone and d'Herelle were not aware of the condition of the patients whose stool samples they were examining.

The choice of hospital is interesting. According to d'Herelle, Campbell Hospital was a pitiful place. The huge open wards were full of faeces-smeared patients. The nurses had only minimal training. There was one rectal thermometer per ward. 'They admitted all the poor souls, the pariahs and casteless. Most of them were in a miserable condition, and many of them had been collected from the street, dying.' That suited d'Herelle to a T. For his study, he preferred this hole to the hospital for British sahibs. He was convinced that under the circumstances on the wards, the disease behaved naturally, making it the right place to understand the ins and outs of cholera and the way bacteriophages work.

For d'Herelle, it was clear that doing experiments in pristine labs on animals that only contracted cholera with a great deal of coaxing was something for mentally lazy cowards who were afraid of raw reality and the dangers of a pitiful hospital for epidemics. The only thing that could be achieved by artificially creating an illness, he thought, was 'a bacteriology, an immunology and an epidemiology for lab animals'. Gloating, he pronounced that if he needed to be treated for cholera in Calcutta, he would pick the hell of Campbell – because there the healing phages spread out among the patients. In the rich people's hospital, hygienic conditions prevented this from happening. As evidence, he cited the fact that in 1926 the death rate from cholera at Campbell Hospital was 27 per cent, while it was 86 per cent at Calcutta's European hospital.

The trio of researchers left the metropolis to tackle the misery in a remote area of the Punjab. Cholera had broken out in the village of Kasur on 8 May and went on to ravage neighbouring villages. They observed the disease as it spread, took water samples from wells and tested them for cholera agents and phages. They discovered that the only villages attacked by the epidemic were those in which they didn't detect any bacteriophages in the wells. The other villages appeared to be protected, just as d'Herelle had predicted.

The three scientists then carried out tests for prevention and treatment, often meeting resistance from villagers, who seemed to reject almost any measure offered by the colonial health authorities. On 20 August the epidemic broke out in Kot Anderson, a village with 800 inhabitants. In four days, 20 people had contracted cholera and 9 of them died. On the afternoon of the 24th, the researchers poured 40 cc of phage solution into the village's five wells. Despite this, there were 9 more cases of cholera in the following three days and 4 of them ended in death. The sceptical research team interrogated the villagers, who admitted that they had pretended to look for a lost ring, drained the main well and then used fresh water. The scientists treated the well water with phages again. It is reported that the epidemic was over one day later. From then on, they stayed one step ahead of the stubborn villagers and contaminated auxiliary wells with permanganate, a substance that turns water purple.

In some settlements, Malone carried out direct treatment tests with willing patients. The large number of rebellious patients served as the untreated control group. Of the 240 control group patients, 60 per cent died, while only 8.5 per cent of the 70 patients who were treated died. An American medical journal commented on the triumph: 'If these statements are confirmed, this will certainly be one of the greatest conquests of bacteriology.'[46]

## One million subjects

The seven-month blitzkrieg waged against cholera had been a success for d'Herelle, and that was all he wanted. The British government invited the pioneer to return the following year, but d'Herelle declined the offer in order to accept an appointment to Yale. He recommended that the authorities employ Igor Asheshov, a Yugoslavian researcher, who subsequently began large-scale use of phages in Patna in Bihar province.[47]

Thanks to Lieutenant Colonel Morison, however, Assam province became the centre of the phage campaign. Morison had lived there since October 1927 as director of the King Edward VII Memorial Pasteur Institute in Shillong. In the institute's report of the same year, he proclaimed: 'The studies by d'Herelle and Major Malone suggest potential, but can only be tested on a large scale. This we do.'[48] Only two years later, 130,839 vials with cholera phages were distributed in Assam, and by 1935 the number had soared to 1,020,000 doses per year. Apparently Morison was dedicated to eradicating cholera with the help of phages in India, his second home. In 1930, the *Indian Medical Gazette* reported: 'Thanks to Col. Morison's energy, bacteriophage treatment and prophylaxis has become almost the rule in bacillary dysentery and cholera outbreaks in the tea gardens in Assam.'[49]

Despite his enthusiasm, Morison remained detached from d'Herelle's announcements of victory. The results of the initial experiments weren't enough to convince him. He felt that large-scale experiments had to be carried out, and he organized them.

Large scale is an understatement. The experiments were gigantic. It was a field trial involving over a million participants.[50]

Since Morison understood the use of phages to be simple and harmless – in keeping with tradition, he had already taken large doses of them himself – he distributed them extensively to the village elders. They were to see to it that the remedy was swallowed as soon as symptoms that could be assumed to be caused by cholera appeared – without waiting for a doctor to come. Would that finally put an end to cholera? Would the fatal epidemics soon be nipped in the bud?

Morison chose two regions in Assam as the setting for his experiment. The area is situated in the eastern tip of what is now India. Several areas, including the test districts of Nowgong and Habiganj, were notorious for their cholera epidemics that regularly washed over the land in spring and autumn. The reason for the epidemic that claimed 10,000 victims in bad years in Assam, with its 7 million inhabitants, was the miserable hygiene in the villages. The people simply deposited their faeces on the flat land, and the rivers swept the excrement away and distributed it. In Nowgong, the Kalang River, with 560,000 people dwelling on its banks, was the executor. In spring, melted ice and snow from the Himalayas would swell the river and flush the stinking mudflats outside the villages. If there was a case of cholera upstream, an epidemic broke out. In autumn, the fatal scenario repeated itself when the monsoon rains subsided and the rivers that sank back into their beds distributed the scattered, stinking excrement a second time.

In the Nowgong field trial, all other regular measures for fighting cholera were stopped, starting in 1929. Vaccinations were no longer administered and wells stopped being disinfected. Instead, vials with cocktails made of various cholera phages were distributed. The Habiganj district was comparable to Nowgong with respect to its location, population and settlement along a river. In Habiganj, however, the local authorities were sceptical about the experiment as were several health functionaries, who didn't want to stop immunizing the people and feared that it would detract from their efforts to eliminate the deplorable hygienic conditions. By default, Habiganj became the control region for the experiment in Nowgong.

It didn't take long for Morison to obtain the first positive results. Despite the epidemics raging in neighbouring villages, Nowgong was spared. Cholera had never taken so few lives. Habiganj, however, was ravaged by the epidemic as usual, until the authorities called for a phage regimen in 1932. The next year cholera claimed the lives of only 35 people in Habiganj, while in the neighbouring district of Sunamgunj, situated at the same river delta, the epidemic that washed downstream killed 1576 people.

In addition to the large-scale study, small individually controlled experiments were carried out in various villages. In some of these experiments, Morison and his team achieved better results than cholera luminary Sir Leonard Rogers obtained in a special department of a prestigious hospital in Calcutta. And all Morison did was distribute phages.[51]

As the *Indian Medical Gazette* reported, Morison's verve led to phages gaining the same significance as vaccination in Assam in just a few years. In 1932, 108,000 people were vaccinated and 191,000 vials of phages were administered. The studies in Assam continued for several years. This gigantic field trial was influenced by so many factors that it was difficult to prove definitively that phages had been the key to the outcome.

## Late satisfaction

Of course, today nobody would even dream of testing a new preparation in this kind of large-scale study. No monitoring authority in the world would allow it. It would be possible to carry out small, compact treatment studies, such as those undertaken in hospitals by Morison, Malone and others.[52] But the idea of dumping phages in wells to protect entire villages from a budding epidemic, as d'Herelle in particular argued for? Never.

At least that seems to be a logical conclusion. Yet exactly this scenario is proposed by researchers in two fascinating publications that appeared in 2005.[53] Shah Faruque and his colleagues work at the International Centre for Diarrhoeal Disease Research in Dhaka, the capital of Bangladesh. The institute is known as the 'Cholera Hospital' throughout the country. Each year over 100,000 patients are treated for diarrhoeal diseases here. Just as it was when d'Herelle was carrying out his studies, in Bangladesh, India's neighbour, cholera epidemics occur twice a year, in spring and autumn. Scientists still do not understand exactly how this cycle is developed. It is considered to be certain that cholera agents are aquatic beings, because they live in rivers and swamps. People ingest the bacteria via drinking water and they make some of them sick. The agents multiply massively in their intestines and return to their surroundings through diarrhoea. There they infect new victims. The epidemic begins and after a few weeks it becomes weaker again.

Since d'Herelle, no researcher would have come up with the daring notion that phages could play a role in this interplay – until Shah Faruque and his colleagues made a discovery that aroused their curiosity. When they examined water samples from three rivers, they found either *Vibrio cholerae*, the cholera agent, or phages that were active against it, but rarely the two together. Could there be a connection there? Faruque then compared the number of cholera cases in the surrounding areas with the occurrence of phages and bacteria in the rivers. And lo and behold, whenever there was an epidemic, he found a high number of agents and a low number of phages in the water. After a few weeks, the number of phages in the rivers rose and the number of bacteria dropped and so, in turn, did the number of cholera patients in the city. Faruque concluded that phages play an important role in ending the epidemic.

The scientists didn't stop at this new piece of information, however. During an epidemic in Dhaka in August 2004, they examined patients' stools for cholera phages. They made another interesting discovery: at first, they hardly detected anything. But the longer the epidemic lasted, the more patients showed the phages. Starting then, the concentration of cholera agents in the rivers was greatly reduced and the epidemic began to subside. Faruque's group postulates that an epidemic proceeds as follows: somewhere in a river, the cholera bacteria get the opportunity to multiply after the concentration of phages has dropped, for instance because the viruses are washed away by heavy rains. The people who collect water or bathe in the river at this place contract cholera and become the origin of the epidemic. The larger bacteria population in the river offers the decimated phage population the opportunity to recover, which reduces the number of bacteria in the water. At the same time, more and more people who have been infected by the river water not only ingest cholera bacteria, but phages as well. In these patients, the phages multiply and return to the water along with their excrement. The balance is adjusted in favour of the phages and the epidemic comes to a halt.

Critics of this theory will now raise the question: 'How can it be that the people who ingest both bacteria and phages in the later phases of the epidemic get sick anyway?' Faruque suspects that there are places in the narrow, winding intestinal tract where cholera bacteria can multiply without being bothered by phages, maybe in the area of the wall of the small intestine, where large numbers of intestinal villi offer a refuge. On the other hand, in the more accessible central area of the intestine, phages are able to multiply. In accordance with this notion, Faruque's team found both high concentrations of phages and cholera bacteria in the stools of many patients. Apparently, simultaneously ingesting cholera bacteria and phages didn't protect these patients from the disease. Yet Faruque assumes that, in many cases, the viruses can protect people from becoming infected, if their superiority in the water that someone drinks is great enough to counteract the cholera bacteria.

These investigations show that d'Herelle's ideas about the course of a cholera epidemic were amazingly farsighted. 'I have no doubt that Félix d'Herelle had the same concept about the role of phages as we have now', says Faruque. 'Amazingly, he thought much ahead of the time.' You might say that he had an ecological view of the disease. He wasn't only interested in how the agents behaved in people, but took other factors into consideration as well. These days he would not be alone in thinking this way. For many infectious diseases such as malaria, the significance of environmental factors in their distribution has long since been acknowledged. And in the case of cholera, d'Herelle's contemporaries also noticed that the aquatic lifestyle of the agents had a major impact on the nature of the epidemic. Yet apparently, d'Herelle's inclusion of phages in the equation meant that he was just ahead of the others – until now. Faruque and his colleagues, who are following the same trail over 70 years later, make no mention of d'Herelle in their publications whatsoever. But they do make a suggestion that would probably have kept him from feeling too put out: 'It seems apparent that cholera phages might be used as biological control agents to interrupt epidemics before they run their natural course.'

## Big money and hair-raising promises

At the same time that Morison was fighting cholera in India, a new phage era began in Europe and the US. Previously, researchers had cultivated phages themselves and used them to treat patients. Now industry entered the arena. In the late 1920s, commercial products appeared on the market. Phage therapy became big business.

In Germany, the German Bacteriophage Society began selling dried phages in tablet form in 1927. Chemists sold pills with staph phages to treat furunculosis (a skin condition characterized by multiple boils) and *E. coli* phages to treat stomach infections. Even then, some doctors wondered whether large enough amounts of staph phages could make their way from the intestines to the boils on the skin.

Antipiol, a company operating out of Berlin-Halensee, offered its product Enterofagos. The 2-cc vials were filled with 'intestinal bacteriophages with a polyvalent effect', said to be effective for typhoid fever, paratyphoid fever, all types of diarrhoea, enteritis, colitis, bacillary dysentery and gastroenterocolitis. At least that's what the advert promised in *Wiener Klinische Wochenschrift*, a weekly medical journal published in Vienna. Medico-Biological Laboratories in London sold a product that went by the same name. In its brochure, the company promised even more than its German counterparts had. Its Enterofagos purportedly cured various infections, as well as hives, eczema and herpes. These are all illnesses that aren't caused by bacteria – which means that phages won't do

**Figure 3.11** Advert for Enterofagos

a thing for them. These absurd promises on the part of industry began to tarnish the reputation of phage therapy.[54]

D'Herelle wanted to prevent such pie-in-the-sky promises of recovery when he consented to the founding of a French phage company in 1928. Since he claimed that he wasn't interested in making a profit, the established pharmaceutical firm Robert et Carriere was to market the products manufactured by the new laboratory. D'Herelle invested the profits in research. He considered himself to be the 'guardian of phage therapy' and insisted to Robert et Carriere that he had a right to veto advertising. Yet barely a year later the advertisers made such exaggerated claims behind his back that d'Herelle feared for his reputation. A legal battle ensued that was fought over a period of years but did not stop the production of the Laboratoire du Bactériophage. D'Herelle's drugs for gastrointestinal infections (Bacté-intesti-phage) or furuncles (Bacté-staphy-phage) spread around the world, as far away as South America and the US.[55]

On the gigantic US market, however, several American companies sensed that there was money to be made in phages. In the early 1930s, at least four big manufacturers had their irons in the fire: Eli Lilly, Swan-Myers of Abbott Laboratories, E. R. Squibb and Sons, which now belongs to Bristol-Myers Squibb, and Parke, Davis and Company, now a part of Pfizer. All four suppliers focused primarily on *Staphylococcus* phages. It appears that the American doctors preferred to administer phages for furuncles, carbuncles and other boils.[56]

## Toxic prima donnas

Despite the offensive by the pharmaceutical industry, phages were only used in a certain percentage of practices. There were a number of reasons for this. For instance, it remained unclear what these enigmatic bacteriophages actually were. Were they a live virus, as d'Herelle claimed? Or were they an inanimate enzyme, as others postulated? As long as these questions remained unanswered, many doctors weren't willing to use the therapy. What was it exactly that they were injecting their patients with? And the fact that the opponents' dispute about the nature of phages was particularly vehement didn't lend credence to the treatment method as far as many physicians were concerned.

The conflict had already arisen in the early years of the young discipline. The first scientist after d'Herelle to study phages disagreed with him – a dangerous deed, as those who had been the object of d'Herelle's rage well knew.

Tamezo Kabeshima, a researcher from Japan, had observed that the phages could withstand a temperature of 70 °C and remained fresh for years without refrigeration. Kabeshima thought that this kind of endurance was more typical of a chemical substance than a living being. The fact that an entire microbe culture dissolved after a few drops of phages were added to it resembled a digestion brought about by some enzymes already known at the time. Kabeshima claimed that it wasn't necessary to postulate a living organism in order to explain the phenomenon that the mysterious substance could be constantly propagated by transferring drops of dissolved bacterial cultures to unaffected ones. Rather, all that was needed was the following chain reaction: a precursor enzyme that exists in every bacterium is activated by the added enzyme – the supposed phages – and can dissolve the next generation of bacteria. The phage, Kabeshima concluded, was none other than a dissolving enzyme.[57]

The attack couldn't have come at a worse time for d'Herelle. After his stay in Indochina in 1920, his workspace had been reduced to the stool in Pozerski's lab and, on top of that, he had the dispute with Calmette to deal with. It got even worse. Jules Bordet of Belgium also began attacking d'Herelle. He agreed that the phages were lifeless enzymes that stemmed from the bacteria themselves. And Bordet wasn't just anybody. He had just been awarded the Nobel Prize, making him a heavyweight who must have made d'Herelle feel threatened, considering his background as a self-taught researcher and outsider.

As if that weren't enough, Bordet unearthed a publication by British bacteriologist Frederick Twort that had already described a similar phenomenon two years earlier than d'Herelle – with the exception that no one had paid any attention to this paper, which appeared in the middle of the war. In an article dated 26 March 1921, Bordet wrote: 'Without meaning to diminish the significance of d'Herelle's observations, we consider it to be our obligation to recognize Twort's indisputable priority in this question.'[58] A lack of funding had prevented Twort from pursuing his discovery at the time and, in his conclusions, he had also remained rather vague. Was it a virus or an enzyme? Twort's cautious opinion was that either one could be possible. Still, Twort was in fact the one who discovered the phenomenon on which d'Herelle's hard-fought and belated glory was based. What mattered here was no longer scientific theories, but prestige and honour. Now the squabbling broke out in full force.

D'Herelle, who had been labelled 'hypersensitive' by an acquaintance,[59] promptly reacted to the attacks. He carried out experiments and issued stinging statements to reinforce his view of things. D'Herelle insulted Kabeshima by claiming that his theory harked back to the dark ages of biology, when foolish scientists still believed in the theory of spontaneous generation. When it came to Twort, he said that his bacteriophages and Twort's phenomenon were two different things.[60]

This thinly disguised evasive defence sparked the wrath of André Gratia. A colleague of Bordet, he later became a friend of Twort – and d'Herelle's most stubborn opponent. In an article subtitled 'Final response to Monsieur d'Herelle', Gratia snarled: 'In his experiments with bacteriophages, Twort described essentially all important things and forgot only one thing: to give it a name.' Gratia went to great lengths to force d'Herelle to confess that his discovery and Twort's were identical. In a protracted series of tests, he meticulously reproduced Twort's original experiments. Even Bordet, who was just as critical of d'Herelle, scolded him for wasting his time. In a letter to Twort dated 27 January 1931, Gratia wrote: 'Two days ago I had a quarrel with Doctor Bordet, who said that my work on the identity of your phenomenon and the bacteriophage [of d'Herelle] is futile.'[61]

As the mud-slinging escalated, d'Herelle demanded that two researchers chosen by the two parties should settle the controversial question by carrying out an experiment. It was an odd procedure that conjures up images of archaic duels rather than hard science. At first, Gratia ignored d'Herelle's demand. But d'Herelle wouldn't budge. He filed a court order to force the highly respected *Annales de l'Institut Pasteur* to issue a second call. This appears to be the first – and last – time that this has ever happened in the world of research. Most of the scholars were incensed at the way d'Herelle had violated the taboo.

In the end, Gratia accepted the challenge. His colleague Ernest Renaux of the University of Brussels and Paul Christiaan Flu of the University of Leiden, d'Herelle's second, carried out the necessary experiments. In spring 1932 they issued their verdict: Twort and d'Herelle's phenomena were the same. The bizarre quarrel about the honour of the discovery had been settled: Twort and d'Herelle had happened upon the same phenomenon independently of each other.[62]

Yet this verdict wasn't enough to put a stop to the debate on the nature of bacteriophages, which continued to rage for more than 10 years. In 1939, the pictures of bacteriophages that Helmut Ruska took with the first electron microscope showed that d'Herelle had been right: phages were viruses. However, when applied to his overall theory, things were not quite that simple. He had always insisted that phages were living microbes that parasitize in bacteria the way bacteria do in humans. Later research revealed a different picture. The phages appeared as beings at the border of animacy and inanimacy. Like all viruses, they can't multiply without host cells.

The dispute as to whether phages are animate or inanimate, however, shifted attention away from the essential issues. The questions as to how phages devour their victims, abuse them so they can multiply and kill them in the process all remained unanswered. Had the phage therapists known this, they could have used the viruses more effectively. The prevailing confusion hindered them from using the bacteriophages more successfully. '[Phage therapy] has fallen short of fulfilling this promise because the men who had to use it have not understood it well enough', *Science* stated in 1929.[63]

## Careless researchers, inept companies

The once euphoric mood began to deteriorate. Ironically, it was d'Herelle's unrelenting hammering that made a major contribution to the decline. In 1930 Paul de Kruif wrote: 'He started the most hopeful hue and cry in a quarter of a century of microbe hunting.'[64] The euphoria of the fledgling period of phage therapy encouraged the belief that the good viruses were a panacea, which undermined the judicious use of the remedy. Enthusiastic doctors used phage therapy to treat diseases that had little or no chance of being cured with this treatment in the first place.

However, some doctors kept announcing successful treatment, only to be second-guessed by other doctors soon after. Confusion and frustration were the result. *Science* published a hailstorm of malice directed at d'Herelle and the creative label of his discovery, which had just been praised by the US media. Now d'Herelle was accused of having merely confirmed and popularized Twort's discovery with his 'picturesque' name. Aggressive critics turned a deaf ear to the voices of the more prudent phage therapists. Paul Hauduroy, one of the original phage therapists, clearly warned the medical community in *Presse Médicale* not to overestimate the effect of the viruses. He stated that phages were only effective for treating certain infections, while they were completely useless for the treatment of others, such as pneumococcal infections.[65]

The phage therapy profession's sloppiness was the major reason for the rounds of vehement criticism. The first issue was that very few reports mentioned the exact amounts of the viruses given to the patients. For the most part, the volume administered was reported, if anything, although this did not shed any light on the number of active phages used. This meant that there was always a risk that patients had received a completely ineffective dose. It was a mistake that continued to be made in more recent studies carried out in laboratories in the former Eastern bloc.

Another problem was that some therapists didn't clearly diagnose the disease they wanted to treat. D'Herelle tirelessly stressed that it was essential to make a careful diagnosis and use it to select the phages. Yet less well-trained physicians ignored this advice. In order to select the proper phages, the doctor first had to isolate bacteria from the patient, cultivate them and test the effect of phages on them in a test tube before they could be administered. Since a virus doesn't attack all strains of a species, several phages had to be given simultaneously. Ideally, doctors should have kept an entire arsenal of phages on hand that could be used to mix the right cocktail.

Phage therapist Ward MacNeal described the high demands as follows: '[Cases such as blood poisoning] present opportunity for a genuine fight against impending death, which requires not only the proper bacteriophages accurately adapted to the individual patient by arduous work in the laboratory but also a fearless, intelligent, skilful and tireless devotion on the part of the physician at the bedside.'[66] This is unlikely to change in the future, even if some day modern phage drugs are introduced to the market. The capricious viruses demand that the physician have much more training than the relatively easily used antibiotics do.

While doctors were overwhelmed by the demands posed by phage therapy, drug producers failed as well. When two researchers from Columbia University tested commercial drugs in 1932, they were confronted with a number of problems. They found preservatives in a product manufactured by Eli Lilly that greatly restricted the potency of the phages. A second drug produced by Eli Lilly proved to be completely ineffective. Squibb, its competitor, took two attempts to produce usable phages. Disillusioned, the researchers wrote: 'If the standard varies so much from time to time, how is the physician to know whether he is using a powerful preparation or merely a tube of broth?' A colleague who had carried out tests that were equally crushing presented the bitter result:

> The reports from these various sources regarding the results obtained with such [phage] preparations are far from uniform. Among the reasons for such discordant results is the fact that a method leading to the preparation of a potent bacteriophage is not generally followed. Thus, when the result of therapeutic application is disappointing, the basic principle is assumed to be faulty.[67]

## A strong headwind

And that's exactly what happened. The *Journal of the American Medical Association* was centre stage for the polemic against phage therapy. In 1934, it published an extensive report about the method.[68] It was commissioned by the US Council on Pharmacy and Chemistry, whose purpose was to protect the public from useless or dangerous drugs. At that time, there was no institutionalized approval authority.

Authors Monroe Eaton and Stanhope Bayne-Jones studied their colleagues' research results and came to a sobering conclusion: apart from the use of phages in staph and bladder infections, there were no convincing results. In nearly all other areas of use, there was an equal number of negative and positive reports. Their dissatisfaction was increased because the nature of phages had still not been clarified.

The major grievance, however, was the quality of the studies, a fully justified point, since nearly all had severe flaws. Many investigations involved groups of subjects that were too small to draw any final conclusions – 15 recovered patients out of a group of 20 patients with dysentery made for a tidy result, but wasn't statistically significant. In the small number of patients, healing could have come about spontaneously in the majority of them.

Also, in a good number of studies, it wasn't certain whether instead of phages, another component of the preparation was responsible for the observed effect. The substance administered or injected into patients was a mixture of phages, bacterial debris and broth proteins, since at the time it wasn't possible to purify the viruses. Eaton and Bayne-Jones suspected that the crude mixture stimulated the immune system, which led to a recovery that was independent of phages – a hypothesis that couldn't be denied and which can in fact play a role.

Roy Fisk of the Los Angeles County Hospital demonstrated what all researchers would have had to do in order to separate the therapeutic effect of phages and the other components of the mixture given to patients. He added extra trials to his series of experiments, referred to as 'controls'. Fisk used a normal phage solution to treat one group of mice infected with typhoid fever. He injected a second group of sick mice with a phage solution he had heated to 70 °C in order to destroy the phages. The result: only the mice treated with the non-heated phage solution survived, while the other mice died. This was an indication that phages were in fact responsible for recovery.[69]

While much of the criticism of the haphazard nature of earlier investigations is justified, it must be remembered that the organization of this type of controlled study needs to be thoroughly thought out and is expensive when it involves developing drugs for humans. The tests often involve hundreds of subjects who need to have similar symptomatologies, ages and states of health for the sake of comparison. Today, clinical studies are one of the most expensive phases in the development of a drug. A standard test involves testing the new medication against the effect of a placebo or an older method. If possible, neither the doctors nor the patients should know which patient is receiving what treatment during the study. Scientists refer to this as a 'double blind study'. The first clinical experiment that met this criterion was not carried out until 1946.[70] Even today, the procedure gives rise to ethical questions such as whether a drug may be withheld from a patient if it is assumed to be more effective.

Doctors had heated discussions about this dilemma even then. Should Morison withhold the phages from the cholera patients in India in the villages where 60 per cent of them died if they weren't treated? Sinclair Lewis succeeded in making an uncannily accurate prophecy in his novel. Arrowsmith, the protagonist, is presented with his first opportunity to use the new remedy during a plague epidemic on a Caribbean island. He goes there and decides to treat only half the sick people in order to procure irrefutable evidence that the treatment is effective. Yet when his wife, who had accompanied him on the trip, dies of the plague, he changes his plan and distributes phages to everyone.

Reality imitated fiction. During the phage experiments carried out in the Indian province of Assam, a similar problem emerged. Lieutenant Colonel Morison, who had been cautiously optimistic, retired in 1934. The experiments in Nowgong and Habiganj were continued. However, it was hard enough for researchers to make valid conclusions in controlled clinical studies, and in the field it was even trickier. This is exactly what Colonel L. A. P. Anderson, Morison's successor, had to report to the cholera committee of the Indian Research Fund Association, which had funded the study.

Since phages had been used in Nowgong, the area had been spared the usual epidemics, but in Habiganj, success wasn't quite so straightforward. Why was there a discrepancy? There must have been a difference between the two regions that made comparison impossible. In addition, phage therapy had become a victim of its own popularity:

> One factor has seriously militated against the success of this experiment and is probably mainly responsible for the absence of conclusive results one way or the other. This is the failure to confine the use of bacteriophage strictly to the experimental areas ... The use of bacteriophage was unfortunately permitted elsewhere ... and this measure has become so popular in recent years that a very considerable quantity is used in every district in Assam during the cholera season. This can only be described as disastrous from the point of view of the experiment.

Anderson saw no way to prohibit the use of phages in other districts and, with a heavy heart, he concluded: 'As regards the prevention of cholera the results of seven years' experiment cannot unfortunately be regarded as conclusive, though the weight of evidence, it is believed, is in favour of this measure.' The members of the committee agreed. They also thought that the indications spoke for the efficacy of phage therapy, but that the experiment could not be saved. It was called off.[71]

Despite the bitter verdict, phage therapy was continued in Assam. In 1938, the Pasteur Institute in Shillong continued to produce 400,000 doses and defiantly announced that the demand from other provinces had risen, where the merits of therapy had apparently been acknowledged. In a carefully controlled study among 1369 patients at Campbell Hospital, bacteriologist C. L. Pasricha found that the mortality rate of patients who had been treated with phages was 13.5 per cent, while it was 16.6 per cent for the control patients. If he included in the statistics only patients in whom cholera agents had been detected, the ratio changed in favour of phages, to a mortality rate of 8.3 per cent as compared to 17.8 per cent. In the distant mother country, this was all considered to be hopeful by *The Lancet*, which reported: 'Thus, the evidence is favourable to the continued use of phage.'[72]

Nor did the US Council on Pharmacy and Chemistry forsake the hope that phage therapy had raised, despite all the complications. After the first report issued in 1934, it commissioned another review in 1941. Ironically, this report also gave a positive assessment of the Indian cholera studies, despite the fact that they had been called off.

In other respects, however, the report, written by Albert Paul Krueger and E. Jane Scribner, came to the same conclusion as the 1934 report: a satisfactorily demonstrated effect could only be observed in staph infections. Otherwise, it was the same old thing. The majority of the studies were of poor quality, although there were some exceptions. Yet the extent of the lack of information about phages at the time is demonstrated by the fact that Krueger and Scribner considered only one thing to be clearly proven: 'Phage is a protein.' They didn't believe it was a virus. Despite all the criticism, this report also concluded with the words:

> Although it is admittedly very difficult to arrive at definite conclusions regarding the efficacy of any therapeutic agent used for the treatment of certain diseases, the accumulated clinical data on phage are in some instances highly suggestive and warrant the continuation of further studies under thoroughly controlled conditions.[73]

The irresistibly simple idea of fighting bacteria with their natural enemies had doctors firmly in its grip.

# 4

## the renaissance of phages during the war

On 1 September 1939, Adolf Hitler's invasion of Poland started the Second World War. Six years later, an estimated 55 million people had lost their lives on the battlefields, in air raids and concentration camps, and they died of injuries, starvation and epidemics.

The war had loomed for years before it actually started and armed forces of a number of countries had made provisions for it. In addition to producing tanks, aeroplanes and cannons, the military also prepared on the medical front. Military doctors quickly realized that they were poorly equipped to deal with potential infections on the battlefield. With few exceptions, such as the antitoxin for tetanus, their arsenal was dangerously empty.[1] They were desperate for anything that would get sick soldiers back on their feet – even the tiny microbes with the tarnished reputation. Suddenly money was pumped into phage therapy again.

The research offensive paid off. Finally, scientists published reliable studies, which applied the knowledge gained earlier about the battle between phages and bacteria. They produced results that continue to be relevant today. Yet how worthwhile would all this progress be? In several labs in the UK and the US, other scientists were working on a drug that appeared to have what it takes to overshadow every known remedy – including phage therapy.

## German rashness

It was primarily German and US military doctors who wanted to send phages, the natural enemies of bacteria, into battle again. While the Americans consistently invested their energy in research, the German army rashly leapt into large-scale production.

Time was running out. Dysentery in particular plagued the German military. In the First World War, 155,000 German soldiers fell victim to it and it killed 8600 of them. The epidemic, which was transmitted through excrement and contaminated flies, could be contained by proper hygiene, but this was almost impossible to manage at the front. Conventional treatment involved confining patients to bed with a heating pad, putting them on a diet of apples and giving them castor oil to flush out bacteria and toxins. In addition to phage therapy, doctors put their faith in antiserums and vaccinations, although their efficacy was widely debated. On top of that, in 1932, Gerhard Domagk of IG Farben, a German chemical company, had discovered that the drug prontosil was effective in killing bacteria. Prontosil was to some extent helpful in treating dysentery and was used in the war. It was the first drug in the class of sulphonamides and continues to be used today to treat several kinds of infections.[2]

Dysentery had already thinned the ranks of the German army during the attack on Poland. Medical officer Professor R. Gantenberg was the head of reserve military hospital 101 in Berlin at the time and soon had to treat the first cases of dysentery:

> The severely ill patients presented an impressive picture of the advanced stage of dysentery: a severely wasted, wan appearance, extreme emaciation, their eyes set deep in their sockets ... Their cheeks were hollow and sunken in, general dehydration so severe

that their skin, especially on the extremities, could be lifted in tall folds and remained that way. The constant tortuous painful urge to defecate ... forced the patients to sit on the toilet up to 40 times a day. In many patients there was a continuous thin stream of intestinal matter ... In the advanced stage the patients had violent bouts of abdominal pain.[3]

For these particularly severe cases, Gantenberg called on polyfagin, a brand-new phage drug that the highly respected German pharmaceutical company Behringwerke had launched in July 1939, just in time for the outbreak of war. As well as polyfagin for dysentery, there was a variant of the same drug for paratyphoid fever. The package insert pointed out that if the drug was administered orally, the sensitive nature of phages required stomach acid to be neutralized. It also advised that 'the sudden massive flooding of the body with bacteriophages is desirable'. Perhaps future endeavours of the German army were behind the insert's warning that 'in hot climates care should be taken that the temperature of the contents does not rise above blood temperature'.[4]

Medical officer Gantenberg was quite impressed by the success of polyfagin treatment. There was a steady stream of press reports about the use of phages, with both positive and negative outcomes. They were usually printed in the journal *Der Deutsche Militärarzt* (The German Military Physician), which also focused on issues concerning 'racial hygiene' in the East and offered advice for the doctor at the front, in articles with titles such as: 'Should the fresh bullet wound to the brain be stitched?'[5]

Phages were used most extensively in Eastern and Southern Europe and in North Africa. According to estimates of several division doctors, 6–10 per cent of all German soldiers in the Soviet Union contracted dysentery in 1941.[6] Military doctors hoped that phages would not only bring about faster recovery but also quickly contain the marauding bugs. Phages appeared to rid patients' intestines of dysentery bugs and in turn stop the distribution of bacteria along with the danger of infection.

In mid-June 1940, dysentery broke out in a POW camp and quickly attacked the German guards. Medical officers Franz Klose and Wilhelm Schröer administered phages and observed an impressive preventive effect. Due to an acute shortage of phages, only 1522 members of the guard and the 251 POWs who worked in the kitchen were given the drug. The result was that the majority of the Germans were spared, while dysentery continued to rage among the unprotected prisoners until October.[7]

The need for drugs to treat dysentery was so great that other companies also began production. The serum institute of Anhalt in Dessau, Germany, for example, manufactured a phage preparation called Asid and sold it to the German armed forces.[8] The German army wasn't interested in the phages produced by d'Herelle's Laboratoire du Bactériophage, however. After invading Paris, the German Wehrmacht monitored the company, but didn't plunder its storehouses.[9] Apparently the Germans only trusted their own products. This turned out to be poor judgement. When the troops of the Axis powers were pushed back at El Alamein, the Allies seized large amounts of polyfagin in 50–500-cc bottles. Military doctors who had been taken prisoner told the Allies that the Wehrmacht routinely used phage therapy to treat dysentery in Africa.[10]

In an ironic twist, pragmatic British doctors tested the captured polyfagin on German POWs. The camp, which was neatly divided up into cages, was the perfect setting for an experiment. While the test results showed that the period of sickness was somewhat reduced, British military doctors weren't particularly impressed. The investigation also revealed that the preparation only had a mediocre effect in the test tube. As it turned out, German manufacturers had been fighting the same problems as their American counterparts in the early 1930s, as described in Chapter 3.

J. Jadin and R. Resseler, two Belgian colonial doctors, came to the same conclusion about polyfagin. The two physicians struggled against dysentery in the Belgian colonies of Africa, present-day Rwanda, Burundi and Congo, where, in 1943 and 1944, epidemics claimed the lives of thousands of Africans. In those times of chaos and need, the new sulphonamides didn't reach the depths of Africa. This prompted Jadin, Resseler and their African staff to cultivate huge amounts of phages in the Rwandan city of Astrida in order to stop the large number of deaths. In 1944, they produced 1100 l of phage solution in tiny 600-ml flasks, which were transported through the jungle as quickly as possible and administered to thousands of inhabitants. The team continually adapted its preparations to the local microbes and achieved successes comparable to those of sulphonamides. This was not the case with the phages plundered from the German army. According to a report submitted by the two doctors, large batches of these phages were shipped off to the Belgian colonies. Their efficacy was poor, however.[11]

The chiefs of the German army's medical staff were not satisfied by the erratic results either. In the *Ruhr-Merkblatt* instruction pamphlet of 1941, it was reported: 'Neither the therapeutic nor the prophylactic efficacy of the phages has been proven with certainty in the case of bacillary dysentery.' The emergence of sulphonamides, which had also been tested in the horrible human experiments in concentration camps, pushed phages aside. The *Ruhr-Merkblatt* of 1944 noted: 'The use of bacteriophages has not proven to be effective.'[12]

## American meticulousness

The Americans took a different tack. Instead of sending in the poorly researched phages, they launched a series of meticulous experiments that soon produced promising results.

The leader of the research offensive was the Committee on Medical Research (CMR) of the National Research Council (NRC). This central committee advised the US government on scientific issues and was assigned the task of keeping the country prepared for war. In 1942, representatives of the CMR visited Morris Rakieten, an employee and friend of d'Herelle while he was a professor at Yale from 1928 to 1932. The two researchers had kept in touch. In a letter to d'Herelle, Rakieten bitterly complained: 'Even *The Lancet* and the *British Medical Times* have had editorials emphasizing the hopeful possibility of phage therapy in the treatment of dysentery in the Middle East and Far East. Both of us should be working in these areas now showing people how to prepare them and use them.'[13]

Rakieten told d'Herelle that the visitors from the CMR wanted to hear everything about phage therapy, but they knew 'desperately little' about the subject. Rakieten presented committee members with studies that he, d'Herelle and Morison had carried out and convinced them of the potential of phage research. Later, the CMR supported renowned scientist René Dubos of Harvard in his animal experiments with phages. In his work, he made an experimental breakthrough, which refuted several of the arguments put forward by phage opponents. Dubos injected dysentery bacteria directly into the brains of 24 mice. Six of the eight mice that Dubos injected with phages were rescued. Untreated mice died from the microbe invasion within two to four days. His achievement was not saving the lives of the mice, but the perfect experimental design, which left practically no room for misinterpretation.

Dubos injected several infected mice with broth without phages. In others, he administered phages that he had heated, killing them in the process. These mice died. This showed that recovery depended directly on the phages and not on immune stimulation brought about by the bacterial debris that was injected at the same time, as many critics suspected. After injecting the phages into their abdominal cavity, Dubos even kept track of their number in the bodies of the mice, proving that the viruses in the blood and brain did in fact multiply – and if enough bacterial fodder was available – at a rapid rate.[14]

The champions of phage therapy had always been convinced that phages multiplied as quickly in the infected body as they did in the test tube. But so far no one had been able to prove it, and the critics considered this to be complete and utter rubbish. Hadn't several researchers demonstrated that blood serum in the test tube reduced the potency of phages? And that there must be substances in the blood that completely eliminated the efficacy of phages? Dubos' findings refuted this fundamental objection.

The promising results were confirmed by similar experiments performed by Harry Morton, Enrique Perez-Otero and Frank Engley of the University of Pennsylvania. They invented an elegant trick to expand the arsenal of evidence. First, they too showed that they could use phage injections to cure mice with dysentery. However, they also injected infected mice with phages that in the test tube had had no effect on the dysentery bugs. Lo and behold, phages that didn't match the bugs did not bring about recovery. Nor did they multiply in the blood and they disappeared in the wink of an eye.[15]

These results show two advantages of phage therapy compared to antibiotics: antibiotics are rather dumb molecules, in that only the ingested amount spreads through the body and is continually inactivated by the organism. Sometimes doctors cannot get the needed amount of active substance to the affected body part, even if a huge dose of antibiotic is given. As in the case of Alfred Gertler (Chapter 1), the bacteria get stuck in the bones and treatment fails. And, towards the end of the treatment, when most of the bacteria are dead, antibiotic residue remains in the body, damaging useful bacteria and leading to antibiotic resistance.

In contrast, phages regulate themselves. Their number explodes as long as they arrive at the scene and find a sufficient number of victims. If needed, the drug reproduces by itself in the patient in a chain reaction. If the bacteria that need to be fought disappear, the phages will no longer find any food and are decomposed by the body and excreted.

Like Dubos, Morton and Engley probably worked on behalf of the US defence department. One of their publications contains the revealing sentence: 'The tests were first conducted in November 1942, under conditions which required secrecy.'[16] Publications submitted by other researchers also contained information that pointed to military use. Arthur Schade and Leona Caroline of the Overly Biochemical Research Foundation in New York developed the production of freeze-dried dysentery phage tablets. In their publications of 1943 and 1944, they mentioned potential large-scale preventive or therapeutic administration; however, where this was supposed to take place was kept secret.[17]

In March 1945, Morton and Engley published an updated report on the value of phage therapy that was again authorized by the Council on Pharmacy and Chemistry (Chapter 3). This time it was limited to the use of phages for dysentery – the disease that the war-effort-driven research had made great strides in treating. The two researchers arrived at more optimistic conclusions than their predecessors in 1934 and 1941.

At first, however, things looked bleak. All the previous experiments had been so flawed that no conclusions could be drawn, either positive or negative. The exception was the experiments on animals performed by Morton, Dubos and several others. They had been carefully carried out – and all of them showed a positive effect. Morton and Engley blamed their predecessors for letting phages loose on humans before they had been properly tested in lab experiments: 'Quite illogically, tests on man were made before the dysentery phage was tried on experimental animals ... The next phase in the history of dysentery phage should be carefully planned prophylactic and therapeutic trials on human beings, taking advantage of the knowledge gained from in vivo tests on experimental animals.'[18]

The next phase never happened. It was spring of 1945. For about a year, a new miracle drug had been available for purchase that had leapt to the front pages of newspapers: penicillin. As described in Chapter 2, in the late 1930s, American researchers had rediscovered the substance that Scottish scientist Alexander Fleming had discovered in 1928 and put aside due to technical difficulties. They developed it into a drug that was first employed by the Allied armies and after 6 June 1944, D-Day, was also successfully used in civilian hospitals.

## Taking aim at typhoid fever, an insidious disease

At the beginning, the first antibiotic was hard to come by and it wasn't effective for all infections. There was no cure for typhoid fever, for instance, until chloramphenicol was introduced in 1947. Until this alternative was found, however, there was a short period that gave rise to a series of excellent phage therapy studies. They provide an impression of what was possible when scientists were more careful in their experiments and knew more about the nature of the bacteria.

The typhoid agents are *Salmonella typhi* bacteria that enter a person's intestinal tract through contaminated water or food. Later they make their way into the blood, accompanied by high fever and, in some cases, confusion and delirium. At the time one-fifth of victims did not respond to any remedy. In the early 1940s, Walter Ward of Los Angeles County Hospital began experimenting with a phage therapy for typhoid fever in mice. He used the ever-growing knowledge about *S. typhi*, which included experiments carried out by his colleague Roy Fisk in 1938, described in Chapter 3. For some time it had been known that not all *Salmonella* had the same properties. The typhoid agent was distinguished by the Vi antigen, part of a molecule on its surface that stimulates the immune system of animals or humans to manufacture antibodies. In 1936, two Canadian researchers discovered a phage that apparently targeted *S. typhi* directly at the Vi antigen.[19]

In a series of meticulous experiments that cost a thousand mice their lives, Ward examined the therapeutic ability of the ultra-selective Vi phages and found that they were much more effective than unspecified viruses that also attacked other species of *Salmonella*. By using Vi phages, the mortality rate of Ward's typhoid-infected mice fell from 93 per cent to 6 per cent. Almost more significant was the fact that Ward used Morton and Dubos' meticulous methodology, which guaranteed that phages were responsible for the recovery.

After these preliminary studies, Ward's colleagues at Los Angeles County Hospital began treating typhoid fever patients by giving them infusions of Vi phages. Out of 56 patients who were treated, only 3 died, a success when compared with the usual death rate of 20 per cent at the time. Within one day, the bacteria were no longer present in the survivors' blood. But what attracted the doctors' attention most was the rapid improvement in the overall condition of their patients:

> Within twenty-four to forty-eight hours after bacteriophage therapy, the patient who had been comatose and in the 'typhoid state' or who had demonstrated the characteristic whining, querulous, obstreperous manner amazed everyone by his cheerful, grateful, cooperative attitude ... Also, patients whose anorexia before treatment was so great as to make forced feeding necessary, afterwards usually asked for food, weakly at first and later, vociferously.

Canadian researcher Jean-Marc Desranleau treated nearly 100 patients in hospitals in Montreal, Quebec City and surrounding areas. He used a cocktail of six different types of Vi phages and reduced the mortality rate to 2 per cent.[20]

Today we know that *S. typhi* entrenches itself in the macrophages, the scavenger cells of the immune system. This poses an interesting question: Do the phages that are on patrol in the blood even get within reach of the hidden bacteria? Or do they only attack the bacteria that romp around just outside the protective immune cells? Today, the answer to this question is more important than ever. This process also plays a major role in tuberculosis, an epidemic that, like AIDS, is gaining ground. Since many tubercle bacilli have become resistant to antibiotics, phage therapy would be a welcome alternative.

As in typhoid fever, the tuberculosis bugs hide in macrophages. Several more recent experiments have shown that phages can reach the immune cells like Trojan Horses, at least in a test tube.[21] However, it isn't clear how efficiently they do this. This is why researchers want to modify phages so they can penetrate macrophages effortlessly and wipe out the tuberculosis bugs there. Two teams have managed to do this in a test tube.[22] The path to using it in treatment, however, will be a long one. For phage therapy, the insidious tuberculosis will probably be one of the toughest nuts to crack.

Pablo Bifani of the Pasteur Institute in Brussels is pursuing an elegant idea. He doesn't want to use the phages to attack the tuberculosis bacteria within macrophages. Instead, he is targeting the bacteria that live just outside the immune cells. Patients in the contagious stage of infection have granulomas (a mass of all kinds of immune cells and eroded tissue) in their lungs. Some of these are connected to the respiratory tract. A lot of tubercle bacilli live in granulomas, both inside and outside the macrophages. The immune system appears to function poorly there and antibiotics can barely penetrate. Every time a patient coughs, he or she spews huge amounts of bacilli into the air. This means that the patient continues to be contagious even after he or she has begun the complicated antibiotic treatment regimen. 'In poorer countries these patients are treated at home despite the fact that they are contagious, because it's cheaper', says Bifani. 'That's why many highly contagious people are out and about there.' Bifani's plan is to have them inhale phages. The viruses travel through the respiratory tract and reach the granulomas, battling the very bacilli that make the patient a source of infection. 'This reduces the amount of time that the patients are contagious', he says. There are other advantages: because the number of bacteria in the body is greatly diminished, the risk that the bugs will build resistances to antibiotics is lowered. Also, the length of treatment is reduced, a huge advantage for a therapy that can last for two years and often fails because many patients simply get fed up with it.

The experiments with the highly specific Vi phages in the 1930s and 40s led to phage typing, an important application that is still in use today. In phage typing, researchers exploit the selective appetite of phages, which often only attack one or just a few strains of a species of bacteria. For instance, one phage only attacks one dangerous strain, while another phage only attacks a harmless strain. If doctors have easy access to specific phages for all the significant strains, then when they are treating a case of a particular disease, they can find out what strain they are dealing with – a dangerous one or a harmless one. If several cases of typhoid fever occur in a city at the same time, epidemiologists can use phage typing to find out whether all the cases were caused by the same strain of *S. typhi*. This would point to a single source of infection, for instance a contaminated well, that could then be identified more easily.

Phage typing is not only used for *S. typhi*, but also for other species such as *S. aureus* or pathogenic *E. coli*. The complicated phage therapy for typhoid fever gradually disappeared when chloramphenicol, which was effective for *S. typhi*, came on the market in 1947. Phage therapy could only continue to claim small niches in a few countries. In France, the successor of d'Herelle's Laboratoire du Bactériophage produced Bacté-staphy-phage, Bacté-intesti-phage and a few other drugs until 1977. They did have the odd adherents, among them physician André Raiga. His petition calling for the company to continue production began with the words: 'I am shattered by the utter perplexity of the patients that for months now have not been able to find ampoules with bacteriophages.' His appeal fell on deaf ears, and the company's phage production was not reinstated.[23]

In Switzerland, Saphal, a tiny pharmaceutical company in Vevey, produced phages well into the 1960s. Coliphagine was used for *E. coli*, Intestiphagine for diarrhoeal diseases, Pyophagine for purulent skin infections and Staphagine for staphylococci. Depending on what they were administered for, they were taken orally in liquid form, injected or used as sprays and salves. The preparations were approved by the Swiss commission for drugs and as such were covered by health insurance there. In Germany, the Dr Heinz Haury chemical factory in Munich sold several of Saphal's phage preparations for DM14 for ten vials. Hermann Glauser, Saphal's owner, had been encouraged to produce phage drugs by French microbiologist Paul Hauduroy, an old friend of d'Herelle. In 1922, he had performed one of the first experiments in treating sick patients with phages. During the Second World War, he was appointed professor at the University of Lausanne, where he became acquainted with Glauser and passed on d'Herelle's legacy to him.[24]

## Loyal till death

Phage pioneer Félix d'Herelle spent the war years in France, a rather unpleasant time. He was by now nearly 70 years old. At the outset of war, he lived in Paris and his cottage in Saint-Mards-en-Othe. When the Germans reached Paris in 1940, the d'Herelle family fled to Vichy, where a friend had found them a flat. In November 1942, the German army occupied this part of France, and d'Herelle, who had retained his Canadian and British dual citizenship, was put under house arrest. Even then he attempted to conduct phage research with employees at the Laboratoire Centrale de Vichy, who were monitoring the quality of Vichy mineral water.[25] In addition, he wrote his scientific doctrine 'The Value of Experiments' and his nearly 800-page memoirs.

In his autobiography, the old pioneer reviewed his rich life, scolded his ignorant opponents and expounded his theories yet again. The unorthodox experimenter also paid homage to the people he admired, including Louis Pasteur – his inspiration – and the eccentric Swiss researcher Alexandre Yersin. When d'Herelle met him in Indochina in 1920, Yersin, the discoverer of the plague bacillus, had built a Swiss chalet complete with an Alpine garden deep in the tropical mountain rain forest and was trying to cultivate the cinchona tree to help the native people in their struggle against malaria. D'Herelle's fondness for Yersin's selfless, nonconformist ways is a statement about his own mindset, which he no doubt considered to be very similar.

D'Herelle revealed even more in his writings, including his zest for travel and adventure, his passion for hunting and love of food, but especially his insatiable curiosity, which gave him no peace. There was always something to discover and he would let nothing stand in his way. In Guatemala, it was the search for the legendary quetzal bird and the Mayan ruins, in India the Taj Mahal, which he sat in front of for an entire day, marvelling at the interplay of the light.

In his tender letters to his daughter Huberte, he revealed himself as a loving father. 'Huberte chérie', the letters began in his sweeping handwriting. From sunny Egypt, where he worked from 1924 to 1926, he wrote that he was sorry it was so cold in wintry France. He complained about the bureaucracy and the tedium it produced, asked how his newborn grandson Théo was doing or announced that he had sent money again.[26]

After the war, d'Herelle achieved a degree of recognition. To celebrate the 30th anniversary of his 1917 publication, researchers at the Pasteur Institute organized a conference where he delivered a lecture. His grandson Claude-Hubert Mazure, one of Huberte's sons, recalls: 'He was very happy that the conference was held and that his work was again the object of attention.' A year later d'Herelle received the important Prix Petit d'Ormoy from the Académie des Sciences. The rules of secrecy associated with the Nobel Prize meant he never learned he had been nominated a total of 30 times.[27] Shortly before his death, his friend André Raiga treated him for a life-threatening illness – with bacteriophages, of course. Raiga was the same scientist who filed a petition to reinstate phage production in France 30 years later. D'Herelle had developed an infection as a complication of an emergency procedure for pancreatic cancer. According to Mazure, the phages were effective for the infection, but the cancer had already progressed so far that d'Herelle died 12 days later, on 22 February 1949.[28]

## Revival

Scientific research on phages was by no means dead, however. It was resurrected far from the sickbed, in basic research where phages became an instrument for solving the mysteries of life. In the 1930s, German physicist Max Delbrück had decided to dedicate his research to biology, because the issues there were more attractive than in the area of physics. Delbrück established an informal group of like-minded scientists who set about decoding the 'secret of life'. They asked: 'How does living matter multiply?' 'How does heredity work?' To answer these questions, they decided to use the simple model of bacteriophages.[29]

It was the birth of molecular biology. Soon after, Delbrück's adherents, who were doing research in many different places, were referred to as the Phage Group. Their research agenda continues to shape the course of biology and medicine even today. The researchers refined the basic method that d'Herelle had used to determine the number of phages in culture media. This allowed them to follow closely the growth cycle of phages, from the infection of the bacteria to the emergence of their offspring. The experiments, which appear to be fairly run of the mill, were the basis for a revolution, because scientists could examine the molecules and mechanisms that played a role in the multiplication of phages. In later research, the viruses helped to answer several fundamental questions that had plagued biologists for years: What does genetic material consist of? What is a gene? Thanks to phages, scientists understood more quickly the code that DNA (deoxyribonucleic acid, the chemical substance that constitutes heredity) uses to store the information for the production of proteins and how the cell duplicates its genetic material.

In the early 1950s, as they investigated these questions, several scientists happened upon a unique phenomenon concerning the phages' selective appetite, which d'Herelle had identified, and which would have momentous consequences. If they added a phage that had grown on a certain strain of *E. coli* bacteria to another strain of the same species of bacteria, it either grew poorly or not at all. What was happening?

A few years later, Swiss biologist Werner Arber unlocked this enigma. The bacteria possess a defence against penetrating viral genetic material, a type of bacterial immune system. They have so-called 'restriction enzymes' that chop up the penetrating DNA but not their own genetic material, which is chemically modified in such a way that defence enzymes cannot cut it up. Those phages that always roam around in the same strain also carry the same modifications in their genetic material. In contrast, the viruses that multiply in foreign bacterial strains carry the wrong modifications and as a result are usually destroyed.[30]

Restriction enzymes revolutionized biology and in turn medicine. They constitute the tools used in every molecular biology lab worldwide to deliberately cut out genes and insert them into any piece of genetic material. Arber's discovery was the birth of genetic engineering, without which the work biologists do today would be inconceivable and we would have neither genetically engineered human insulin for diabetics nor gene therapy.

The discovery of restriction enzymes also provided a convincing explanation as to why many earlier phage therapy experiments had failed. If a doctor used a purchased preparation, which had been cultivated on a bacterial strain other than the one torturing the patient, the penetrating genetic material of the phages was destroyed by the bacteria's immune system. The therapy failed. Only if phages were tested in a test tube with bacteria from the patient could doctors detect this. D'Herelle had proposed this exact method and had even called for a standard procedure, which involved not only testing phages on the bacteria of a patient but also cultivating them on these same bacteria. He had achieved his best results using this approach. Today we know why this was the case – phages could adapt to the restriction enzymes of the bacterium being targeted.

However, for some phages, this type of adaptation phase doesn't seem to be required. In the battle against bacteria that has been waged for millions of years, they have been armed against the immunity of their foes. These phages modify the chemical structure of their DNA so that the restriction enzymes can no longer cut them up. Other resistant phages have largely banned the recognition sites, which each restriction enzyme needs in order to be able to cut, from their genetic material. New investigations carried out by Georgian phage researchers have shown that such protected phages are among the most effective ones in their therapeutic arsenal.

The research carried out by the Phage Group in the 1950s also brought some reconciliation between the once hostile camps – the one camp that, like d'Herelle, claimed that phages were viruses and the other that, like Bordet, postulated that they were merely enzymes from the bacteria themselves. Experiments by French researcher André Lwoff and other scientists revealed that some bacteriophages incorporate themselves in the DNA of a bacterium instead of multiplying immediately. Strains of bacteria that carry such quasi-inherited phages can suddenly produce viruses as a result of certain stimuli. This means that Bordet's view that phages are components of bacteria is to some extent correct.

## The invisible world power

Due to the many successes, it quickly became impossible to keep track of the burgeoning field of 'bacteriophagia'. In 1931 alone, there were approximately 2000 publications.[31] By 1965, Hans-Jürgen Raettig of the Robert Koch Institute in Berlin, who monitored all publications on this subject area, found 11,405 publications. Since then people have lost count. Nor can the number of phages be recorded. Estimates put them at $10^{30}$, totalling some 10 billion times more of these bacteria killers than there are stars in the universe, making them the most numerous organisms in the world.[32]

'There are bacteriophages nearly everywhere', says phage researcher Hans-Wolfgang Ackermann of Laval University in Quebec City. They colonize the human intestinal tract and the skin and have a predilection for living in sewage, lakes and rivers, soil and even in 100 °C salty hot springs, as well as in food. Every day, we ingest millions of viruses. Scientists have tracked down phages in meat, whether fresh, rotten or cooked, beef or chicken. They have also found the viruses in raw milk, mushrooms and lettuce.[33]

There are phages wherever you look, in a huge range of shapes and sizes. Scientists have already identified over 5000 variants under the electron microscope. Most of them look like the prototype, which resembles a lunar module with a crystal head, tail and tail fibres. Others are threads or small balls, rods or amorphous masses.[34] Yet only a small percentage of all types of phages is known to scientists. Microbiologists estimate that they are familiar with only about 1 per cent of all bacteria. The rest of them are difficult to cultivate in a lab because their demands for food or living conditions aren't known, and as long as the bacteria are unknown, their natural enemies will also remain unknown.

'We hardly know anything about bacteriophages', says Ackermann tantalizingly, who has dedicated 40 years' research to them. He still goes to the lab every day, even though he is retired, to help take care of the renowned phage collection that he set up. It's true that the mountains of evidence are dwarfed by what isn't known about phages. Where do the viruses come from, for instance? 'No one knows', Ackermann says. One theory says that the source may be renegade bacterial components. The important role played by phages, as they do their devouring and are devoured themselves, is only slowly being revealed. Using highly sensitive techniques, researchers have recently found amazing amounts of phages in sea water. Up to 100 million phages per millilitre are waiting for their food there. Their number varies widely depending on the rhythm of the seasons and the food available. No doubt they play a central role in the coming and going of plankton, which fulfil a central function in the ecological interplay of the seas as staple food. Phages are the hidden controllers of the oceans.[35]

People sensed the power of bacteriophages long before they knew anything about them. For ages, countless litres of milk have not turned into cheese because bacteria that were supposed to do the job were killed off by viruses. Phages continue to be a problem for today's cheese makers, just as they are for the biotech industry, where they sometimes destroy precious cultures of bacteria that produce drugs.

On 6 May 1944, the destructive potency of phages even made it to the *New York Times*. A newly constructed factory in Puerto Rico belonging to the US government was designed to produce butyl alcohol with the help of bacteria. The butyl alcohol was to be used for the production of rubber, which was key for the war. Phages inadvertently brought into the factory killed the bacteria and paralysed production.[36] Félix d'Herelle would have liked this news. The phages seemed like a dying creature rearing its head for one last time. Ten years earlier, they would have made the headlines of the same newspaper due to their healing powers, but now they were only associated with destruction. The hope that phage remedies had provided for so many people would now remain forgotten for a long time – at least in the West.

# 5

## a parallel universe

What a happy person. He had escaped the gruesome war by a few months and managed to get away from the icy expanses of Siberia. Instead, newly qualified doctor Teimuraz Chanishvili had just started doing research at the Tbilisi Institute for Microbiology, Epidemiology and Bacteriophagia. Things weren't exactly perfect, though: 'The other student who had come to the institute with me was horrible', recalls Chanishvili, who retired as director of the institute in the summer of 2005 aged 81, laughing mischievously. 'We despised each other. She even tattled on me once to the boss when I showed up at work completely exhausted.' And that was no small matter, considering that Elena Makashvili, his boss, had a forceful personality and demanded that her lab workers fulfil their duties.

But this was all better than the alternatives. In June 1945, Chanishvili had graduated from university in the Georgian capital of Tbilisi. Had he finished a few months earlier, he would have had to serve his fatherland in the war. After the war, most of his colleagues had been sent to the most wretched corners of the immense Soviet empire, where the change of seasons only meant the substitution of mud for snow. Chanishvili's charm apparently had the same impact then as it does today. The institute's directors liked him and offered him a job.

The Tbilisi Institute for Microbiology, Epidemiology and Bacteriophagia wasn't just any old institute. It was the venue of heroic achievements. The institute's researchers had supported the desperate battle of the Red Army against the German invaders. In fact, they had even made it possible by winning their battle, the battle against lethal microbes. Here, in the new part of the institute, with its grand staircase and mighty columns, scientists had cultivated phages in circumstances of great hardship in order to save soldiers from dysentery and wound infections. Phage researchers undertook over 170 dangerous trips to fight their battles against bacteria. They travelled to the front in order to teach military doctors how to administer phages or test the efficacy of new preparations.[1]

Yet the institute in Tbilisi concealed a dark secret. Georgiy Georgievitch Eliava, its first director, had done a number of remarkable things. After founding the institute in the 1920s, he made it one of the flagships of the Red Empire. He brought the renowned Félix d'Herelle to Tbilisi and paved the way for phages in the Soviet Union.

And then, in January 1937, he suddenly disappeared.

It is difficult to reconstruct the tragic story of phage pioneer Georgiy Eliava, despite the fact that the institute continues to be one of the most important centres for phage therapy today, where Alfred Gertler found healing for his infected foot (Chapter 1). You can count the documents about Eliava that have come to light over the years on one hand. Nearly everyone who knew him personally died before the taboo on talking openly about Eliava was lifted. What remains are stories about him that have been passed down from his stepdaughter and employees and the recollections of 94-year-old Nina, 'Nunu', Kilasonidze, which she shared with me one day in her Tbilisi apartment. She worked with Eliava during the four years prior to his disappearance and has an uncanny memory for the details of the tragic period.[2]

## A momentous friendship

The roots of the institute that Eliava shaped, and where Chan-ishvili started his career eight years after Eliava's disappearance, go back to the 1920s. It was a troubled time, not only in Georgia, perhaps, but especially there. On 11 February 1921, the Red Army invaded the little Caucasian republic. After three years as an independent state from 1918 to 1921, the successors of the tsars brought the Georgians back under Russian power again. The usual social cleansing took place.[3]

Yet Georgiy Eliava, the young director of the Institute for Microbiology in Tbilisi, kept his position despite his social background. Eliava belonged to the bourgeoisie. He was born to a prominent physician and an aristocratic lady in January 1892 in the village of Sackhere in the foothills of the Caucasus. His great aunt had made a fortune in manganese mines. This was exactly the kind of family that the Communists had trained their sights on. Although Eliava participated in a few assemblies of revolutionary students while he was a student at the University of Odessa, causing him to be expelled, it wasn't enough to outweigh his heritage.[4]

It must have been his skills that kept the director from being dismissed. They had already served him well early on. After being expelled from Odessa, he studied in Geneva from 1912 to 1914.[5] There, Eliava decided to stop majoring in literature, his first passion, and change to medicine. The First World War broke out while he was at home during the term break. Returning to Geneva was out of the question. His knowledge, along with his great aunt's money, allowed him to enrol at the University of Moscow, where he graduated with honours in 1916, at the young age of 24.[6]

Eliava immediately became the head of the bacteriological lab in Trabzon on the Black Sea. One year later he was appointed the director of the microbiological lab in Tbilisi. It was the first lab of this kind in Georgia and was urgently needed. Dysentery and diphtheria were raging, and there were repeated outbreaks of cholera. In 1918, the government of the independent state of Georgia sent Eliava, who was only 26 at the time, to the Pasteur Institute in Paris. In this temple of medicine, he was expected to gain the knowledge needed to set up a modern bacteriological research institute in his homeland. Eliava learned about the production of vaccines and serums and bought equipment.[7]

Shortly after his arrival in Paris, Eliava made a momentous acquaintance, which, like so much in his mysterious life, has become legendary. A year before, while still in Tbilisi, Eliava had made a strange discovery. He was studying cholera and investigated the water of the Mtkvari River,[8] which flows through Tbilisi on the way from northern Turkey to the Caspian Sea. Eliava wondered whether the river could be spreading cholera. One day he found cholera agents in water from the Mtkvari. He was called away to a meeting and left the sample under the microscope. When he returned, the cholera microbes had vanished into thin air, leaving the liquid behind. Eliava tried to find an answer to the phenomenon but came up empty-handed.[9]

A year later, while at the Pasteur Institute, he heard about Félix d'Herelle's discovery, which Emile Roux, the institute's director, had just announced at the Académie des Sciences. Was this the explanation for his experiment with the water from the Mtkvari? Eliava requested permission from Roux to repeat d'Herelle's experiments, which he succeeded in doing. He then supported d'Herelle's theory that phages were living viruses, which was already being debated at the time. Roux sent a telegram with this news to d'Herelle, who was tackling fowl typhoid out in the countryside. D'Herelle rushed to Paris, and in the lobby of the Pasteur Institute, he cried, 'Where is this Georgian?' Eliava appeared, and the two men embraced each other like bosom buddies.

In fact they became close friends. 'D'Herelle and Eliava valued each other greatly and liked each other', wrote Pasteur researcher Edouard Pozerski in his memoirs.[10] D'Herelle had found refuge in Pozerski's lab after his argument with Calmette, and temporary guest privileges were also extended to Eliava. In one of Eliava's few surviving letters, he talks about the 'purity', 'openness' and 'warmth' of their relationship.[11] D'Herelle and Eliava were the epitome of the 'odd couple'. D'Herelle, the mentor, was nearly 50, an unsociable adventurer with many enemies and for whom science was the greatest good. Eliava, the pupil, hadn't yet turned 30, was a talented researcher, but also a playboy, urbane and popular wherever he went.

## A man for legends

In the few surviving pictures of Eliava, we always see an amiable man, with a round, pleasant face and elegantly dressed. He must have been very charismatic. Nunu Kilasonidze, who worked with him many years ago, speaks of him tenderly and full of admiration: 'When he entered a room, all the young ladies fell in love with him right away.' She was in her mid-twenties at the time. His magnetism worked in Paris as well. Elie Wollman's parents worked at the Pasteur Institute, and as a child, he met Eliava. Wollman reports that the wives of the stern Pasteur researchers gossiped that Eliava had made the close acquaintance of many a Parisian lady during his stay.[12]

Surprisingly, Eliava's success with the ladies didn't seem to bother his male colleagues; he got along with everyone. 'He just exuded charm', Wollman recalls. During my visit, Nunu said: 'Whether they were professors or lab technicians, Eliava treated everyone with the same warmth. He liked his employees and spent a lot of time with them. Sometimes he stood at the entrance to the institute in the morning with his lips puckered, waiting for the employees. Then he gave them all a kiss.' Nunu puckered up her mouth in her wrinkled face and showed how the great master used to kiss. 'Was it only the ladies?' 'No, no, he kissed everybody!'

After he returned from Paris in 1921, Eliava married Polish opera singer Amelia Vol-Levitskaya, who was quite a bit older than him. A photo shows the diva in a smart white dress, ready to make her entrance to the Tbilisi opera. Amelia suited him well. He loved opera, literature and dance. His stepdaughter Hanna couldn't imagine that her father was a scientist. In an interview with a Georgian journalist in 1988, the first article to be written about Eliava since his disappearance, Hanna, by then an old woman, recalled: 'He was so full of joy. It seemed completely improbable that he could be constantly obsessed by an idea. At home we had a rule that we weren't allowed to talk about microbiology.'[13]

Eliava was, however, a serious scientist. He appears to have made such a good impression on the luminaries at the Pasteur Institute that they offered him a position in France, his step-daughter later told the journalist. Yet he is said to have answered: 'In France there are many microbiologists, but in Georgia, I'm the only one. Georgia needs me.'[14] In November 1921, Eliava sailed back to the Black Sea port of Batumi. Not only did he turn down a good position in France, he also risked life and limb. Georgia was now under Soviet rule. Many of his fellow Georgians with aristocratic roots left the country.

The courageous Eliava brought back lab material worth F100,000, as well as all kinds of knowledge. He set about establishing the field of microbiology, which had been non-existent in Georgia up to then. In 1923, the Institute for

Microbiology was officially founded, with the 31-year-old Eliava as its administrative and scientific director. A report issued by the institute in 1926 listed a wide array of research projects. Under Eliava's leadership, scientists did research on bacteriology, as well as leukaemia and immunology. According to Nunu Kilasonidze, Eliava carried out dangerous experiments himself: 'There was a special building for plague experiments. Eliava dissected the animals that had been infected with plague bacillus himself.' The report also mentions a lecture given by Eliava called, 'On d'Herelle's phages'.

The institute was the only one in Georgia that produced vaccines and serums, and it did this so successfully that the country became self-sufficient. Despite his youth, Eliava was soon appointed professor of hygiene at the medical school in Tbilisi.[15]

During this period, the Communists allowed the brilliant scientist to work undisturbed. He was unaffected by a wave of social cleansing that took place following revolts in 1924, although hundreds of people were shot and entire villages burned to the ground.[16] Eliava was even sent to the Pasteur Institute again, from 1925 to 1927 and from 1931 to 1932.[17]

It wasn't as if Eliava courted the tolerance of the powers that be by keeping a low profile when it came to his lifestyle. On the contrary, he loved bourgeois pleasures such as horse racing and made no bones about it. 'The institute had an entire stable of horses that were kept for serum production', Kilasonidze told me. 'Eliava also had them run in horse races. But they always came in last, since they were always having blood taken from them.' Yet Eliava, a horse aficionado, didn't stop at the small pleasures. He also indulged in buying a thoroughbred in Paris that he brought back to Georgia.[18] Another time he brought back perfume for the ladies in the lab. Since importing perfume was outlawed, Eliava smuggled it into the country in lab containers.[19]

## The Soviet Union makes an exciting catch

During his third stay in Paris, Eliava stayed as a guest at d'Herelle's country cottage.[20] At the time, d'Herelle was a professor at Yale but was spending the summer in the French countryside. While the appointment at Yale was the most prestigious position he had ever had, things weren't looking good. He had fallen out with the dean and the administration[21] and was up for new adventures. Eliava was hoping for the breakthrough of phage therapy in the Soviet Union, and who would be better to help him than the phage pioneer himself? It wasn't long before Eliava was officially able to offer d'Herelle a position at the Tbilisi institute. Engaging the famous, albeit controversial, star must have been a great coup for the Communists. Now that their power had been established, they were boosting industrial production and building up scientific know-how.

At the time of the invitation, however, the friendship between master and pupil no longer appeared to be so close. Eliava made some critical remarks about his mentor in a letter addressed to Pasteur researcher Edouard Dujardin-Beaumetz. He was upset about an advert for d'Herelle's Laboratoire du Bactériophage. Clearly disappointed, Eliava wrote:

> I admire his ingenious judgment that he had when he observed the new phenomenon of the bacteriophage, but that makes it even more inexcusable that he is using his fame for money and hurting himself in the process. You must have noticed that our relationship, that you ... must have heard about was not able to maintain its original purity, openness and its past warmth ... This discontentment, which I visibly felt, is not due to anything personal (on the contrary, d'Herelle has always behaved absolutely perfectly towards me), but I had the feeling that d'Herelle had changed quite a lot.

He said that d'Herelle had become dogmatic and put his theory above the experimental facts.[22] This is a serious accusation for a scientist. Eliava was not the only one to criticize d'Herelle for making money from his discovery. D'Herelle's grandson Claude-Hubert Mazure, however, stressed that his grandfather never received money from the Laboratoire du Bactériophage.

It has often been asked how d'Herelle could work in Stalin's gloomy empire, of all places. In contrast to the accusation about lining his pockets with his research results, he was later accused of being an enthusiastic Communist.[23] This accusation was fuelled in recent years when several copies of the Russian translation of his book published in 1935 reappeared, which had been kept under wraps from the time of Eliava's disappearance until the collapse of the Soviet Union. The first page features a dedication to Joseph Stalin, with Félix d'Herelle's signature.[24] Was d'Herelle an admirer of the murderous dictator?

D'Herelle's grandson Mazure rejects these ideas. He claims that his grandfather had never been a member of any kind of party. A look at the situation then and d'Herelle's writings supports this point of view. In his memoirs, d'Herelle refers to himself as a 'Socialist', a term he distinguished from the label 'Communist'. At the same time, he 'loathed' politics.[25] According to Mazure, his grandfather didn't know anything about the dedication to Stalin. The book, which d'Herelle wrote in Georgia and Eliava translated into Russian, wasn't published until d'Herelle had returned to France in 1935. The dedication to Stalin was pasted into the book and the signature is a facsimile. Even had he known about it, it doesn't mean that d'Herelle was a Stalinist: 'Back then, this kind of dedication was simply a guarantee that the book would appear', says David Shrayer, a Russian scientist who emigrated to the US. 'Even critical writers like Boris Pasternak had dedications like this one in their books. D'Herelle's dedication is on the low end of the scale. I've seen much more elaborate dedications in other books. It's like a stamp, that's all.'

In fact, the text sounds stilted, made up. The gist is that a researcher's most important task is 'to minimize man's suffering'. It closes with these bombastic words:

> This book summarizes nearly 20 years of research on new paths in medicine. I dedicate it to the one who allows himself to be guided by an unsparing and uncompromising logic of history and thus builds on a fully new foundation. In fulfilling this task he has reached such perfection, that an unbiased observer must acknowledge his deeds. I dedicate this book to comrade STALIN.[26]

One thing is for certain: d'Herelle did not like capitalism. He had experienced the Great Depression during his professorship at Yale and despised what he had seen: 'The very fact that even at the apex of wealth millions of people are without work demonstrates the powerlessness of the capitalistic system to guarantee each person a basic existence.'[27] D'Herelle did, however, acknowledge the positive sides. He praised the good universities and agricultural policies. But he felt that in the long run, the US would not flourish. In his memoirs, the chapter about his stay in America is entitled 'The End of a World'.

D'Herelle never shut his eyes to the social injustices he encountered. On the hacienda in Mexico, he discovered that the owner secretly whipped the Indios who worked there. As a guest in the manor house, d'Herelle had to dig around to discover this abuse. 'The workers doing the planting were slaves', he wrote in his memoirs. D'Herelle wanted to expose these conditions after his departure but, soon after, the Mexican government was overthrown and the conditions, also found at other farms, were said to have improved.[28] Later, when he was in India, he wrote about the caste system: 'The Brahmans are the only people in the history of man who have managed to shape their people such that they can use it for indulging in their own vices to this extent.'[29] He thought the collective farm economy of the Indians in Mexico prior to the Spanish conquest was a model society.[30]

Under the circumstances, d'Herelle felt that the alleged class-less society in the new Russia was at least worth looking into, especially as science was deemed to be important. The comrades wanted it to launch the archaic agrarian country into the industrial age. 'The Soviet rulers have assigned science all the authority that they have stolen from religion', French politician Edouard Herriot wrote at the time.[31] To d'Herelle, a critic of religion and a disciple of science, this message boded well. He wrote: 'The domains of the experiment must extend to every human thing, because under its patronage all hopes of well-being quickly become reality.'[32] This could have come from Lenin himself.

The strongest argument for moving to Georgia, however, was the prospect of bringing about the breakthrough of phage therapy in this huge country. This was clearly behind the Soviets' invitation. Before the revolution in 1917, nothing there warranted the name 'health policies',[33] and, because of this, epidemics still raged in the Soviet Union. Lenin appealed to the citizens of the country to fight the plagues: 'We need to direct all the resolve and experience from the civil war to the war against epidemics.'[34]

Félix d'Herelle had this resolve. He was 60 when he and his wife Marie again boarded a ship for new shores in October 1933. He stayed in Georgia twice, from October 1933 to April 1934 and from November 1934 to May 1935. D'Herelle was treated like a celebrity. His working conditions were good, and he had his own lab staff. It wasn't long before he could publish his book and he became an honorary professor at the University of Tbilisi.[35]

Eliava tirelessly took care of his mentor and his wife. After their arrival in Batumi, Marie wrote in her diary: 'I received beautiful flowers from Eliava.' The next day she, d'Herelle and Eliava attended the festivities of the anniversary of the October revolution. 'It was very nice', she noted. In the following weeks, she encountered a number of things she considered 'nice', especially the many new buildings. Apparently, in Batumi, Tbilisi and elsewhere in the Soviet Union, the establishment of the new state was forging ahead.

The couple saw more of this on several trips with Eliava to Leningrad, Moscow and Baku. D'Herelle's snapshots often show construction sites. In Moscow, the 42 buildings of the Soviet Academy of Science were being constructed for the astronomic sum of Rb225 million. In his usual way, d'Herelle recorded an enormous amount of facts. Elections were taking place: 'Here, everyone is allowed to vote: members of the military as well as foreigners, as soon as they have started working', d'Herelle noted, impressed. Several foreigners were even elected: 'Nevertheless, Stalin was at the top.'[36]

## Dangerous moves for a grandiose vision

Despite the sightseeing tours, phage therapy was the main focus of d'Herelle's stay. In the director's report published to mark the 50th anniversary of the institute in 1974, Irakli Georgadze, who had been d'Herelle's assistant, wrote: 'D'Herelle arrived at the lab at 8:00 am. He worked a lot, and it was difficult for us young assistants to keep up with him. Professor d'Herelle was very clever and had a fantastic technique.' In the introduction to his book published in 1935, d'Herelle wrote:

I am daring to hope that my book can serve as an impetus to find out about new epidemiological perspectives. I see this as being especially desirable considering that in the USSR an era of rebirth and an unprecedented upswing in science has to begin. Biological laboratories are being set up everywhere in our huge country. Here the scientific life is becoming more and more intensive, while in the capitalistic countries there is a rising tendency towards deceleration. Science was condemned to be the first victim of the global economic crisis.[37]

D'Herelle and Eliava made plans to set up a grandiose institute – just for bacteriophages. They wanted it to be the world's phage therapy centre. In addition to extensive research buildings, they planned several hospitals, luxuriously furnished with two-bed rooms with one doctor and two nurses per room. This was a daring idea at the time, and not just in the Soviet Union. 'All of the buildings were supposed to be lined with marble', recalled Kilasonidze, 'and phages were to be used everywhere.'[38]

Eliava had the necessary contacts among high-ranking party bigwigs to realize this dream. One of them was Lavrenti Beria, first secretary of the Central Committee of the Georgian Communist Party and head of the secret police there. Stalin is said to have called him a promising fellow Georgian and brought him to Moscow in 1934.[39] Beria was a powerful acquaintance, but also a dangerous one. Then there was Grigorii Konstantinovich Ordzhonikidze, a Bolshevik veteran and war comrade of Stalin. As minister for heavy industry, he belonged to the inner circle of powerful men in the Soviet Union.[40] Polykarp Mdivani, also a revolutionary from the beginning and the dictator's fellow traveller, completed the list of important contacts.[41]

It appears that Eliava approached Beria first. In Marie d'Herelle's diary, the entry dated 29 November 1934 reads: 'We're going to the Eliavas. He has left in order to meet Beria.' At one of these meetings he is said to have presented Beria with an official petition for his phage institute, which Beria rejected.[42] Then Eliava got Ordzhonikidze involved, making sure that he received a copy of d'Herelle's new publication.[43] In so doing, he opened up direct access to the inner sanctum of government – bypassing Beria in the process. It was a fatal error.

On 2 December 1934, the d'Herelles and Eliava boarded the train at Tbilisi. They were headed for Moscow for important negotiations. Yet even the beginning of the trip seemed to be a bad omen. 'Before it could depart, the locomotive [hit the train] with a mighty bang, which destroyed two of the cars. I was pressed against the edge of the wash basin and received a violent blow. The pain was ghastly. What a great beginning! I already had a little sore on my left leg that was growing! And eczema on my ear!!!', d'Herelle wrote in his notebook.[44] He was tortured by pain during the entire trip. It transpired that he had broken a rib. On top of that, d'Herelle, a chain-smoker, suffered from stubborn bronchitis and pneumonia.[45] Age was taking its toll.

Eliava told him the really bad news a short time later, although d'Herelle probably didn't realize the implications at the time. 'As we departed, Gogi [Eliava's nickname] came to tell me that Kirov had been murdered in Leningrad.' Sergei Mironovich Kirov was the Party leader in Leningrad, considered to be Stalin's best friend and successor.[46] D'Herelle's only comment related to the murder in his notes was, 'Strange'. And indeed it was.

A failed comrade had shot the Party leader. The murder had been strangely easy for him to carry out. At the time of the attack, all the guards had been withdrawn from Kirov's official residence. He was probably murdered on Stalin's orders, although this still remains unconfirmed today.[47]

The dictator wasted no time in using the assassination for his purposes. *Pravda,* mouthpiece of the Communist Party line, declared: 'The enemy did not fire at Kirov personally. No! He fired at the proletarian revolution.'[48] Claiming the murder to be part of a widespread plot against the Soviet leadership, Stalin initiated the Great Purge. This wave of destruction claimed tens of thousands of lives in the next four years and rocked the Soviet Union to its very foundations. The dream of a world phage centre had suddenly dimmed, even if this was lost on d'Herelle at that point.

After all, everything appeared to be going according to plan. On 27 December 1934, d'Herelle was received by the minister of health in Moscow. 'Félix met the people's commissioner for health [Grigorii] Kaminskii, who made the decision about the construction of a laboratory for bacteriophages', Marie d'Herelle wrote in her diary. Kaminskii invited the phage pioneer to continue his work at an institute in Moscow. However, d'Herelle declined, saying the climate wasn't good for his chronic bronchitis. A few months later, in May 1935, the d'Herelles travelled to France for the summer. They were sure that they would return to Tbilisi in the autumn. In a letter to the Party, Eliava had written, 'D'Herelle has already produced two years of scientific work and is eager to continue it.'[49] At home in France, Félix told his grandson Claude-Hubert that he would soon be taking him on holiday to the distant country in the Caucasus.[50]

There were signs, however, that it would turn out differently. Before leaving for France, for example, the show trial for the murder of Kirov took place. And in his notes, d'Herelle wrote that a certain Garloch hadn't shown up for a meeting. Rumour had it that he was afraid to be seen with foreigners because he had already been arrested for meeting some Germans. As early as 1933, Eliava himself had been detained along with 16 Georgians for 'sabotage'. Beria was behind the arrest.[51]

## Goal achieved – everything lost

Despite this, the situation appeared to calm down. There were no more trials after the Kirov trial, and a short time later, on 14 April 1936, the Council of People's Commissars of the USSR promised Rb13 million for the construction of an entire complex for phage therapy, including the hospitals. A gigantic 17 hectares – 42 football fields – were reserved for this purpose in the Tbilisi neighbourhood of Saburtalo. An expansion of the institute into the world centre for phage therapy had already been planned.[52] Everything seemed to be in place.

D'Herelle should have received this good news in Tbilisi, but he wasn't there. After his summer holiday in France, he waited in vain for a new visa from the Soviet Union. Instead, there was only deafening silence.[53]

We don't know how much Eliava knew about what was happening in the USSR, but he couldn't have thought that it was a good omen. Dark clouds were beginning to gather again on the political horizon of the Soviet Union. Under the circumstances, his increasingly poor relationship with Georgia's head of the secret police was a dangerous burden. Much later, Eliava's assistant Makashvili recounted the following incident: one day Beria was suffering from high fever. Eliava and Makashvili were called to his bedside. Eliava took blood from Beria, who joked, 'Don't suck all the blood out of me', to which Eliava responded: 'This is nothing compared to the way you have sucked blood out of the people.'

**Figure 5.1** Félix d'Herelle (centre) in Tbilisi, watched by Georgiy Eliava and his assistant Elena Makashvili

In August 1936, Stalin's henchmen dragged Party veterans Grigorii Zinoviev and Lev Kamenev to court. They had already been sentenced in connection with the murder of Kirov. The rhetoric at the trial left no doubt about the state of the country. 'One cannot find words to fully express one's indignation and disgust. These people have lost the last semblance of humanity. They must be destroyed like carrion that is polluting the pure, bracing air of the land of Soviets, dangerous carrion that may cause the death of our leaders', snarled Party cadre Grigorii Pyatakov. Beria also used threatening words: 'A Communist who demonstrates reconciliation and depraved liberalism in the face of falsehood, no matter what form it appears in, commits the greatest crime against the party, the power of the Soviets and the mother country.'[54] Zinoviev and Kamenev were executed.

Nearly 70 years after this dark period in Russia's history, Nunu Kilasonidze vividly recalls those times. Sitting in the armchair in her living room, she reeled off the names of Eliava's friends who were arrested and then disappeared:

**Figure 5.2** Georgiy Eliava

'Vladimir Jikia, a construction engineer, or the Aganiashvili brothers. One of them was the rector of the Tbilisi University and the other a member of the government of the Georgian Soviet republic, just like Shalva Matikashvili.' Despite the sun shining brightly through the balcony door, the room suddenly seemed to darken. Several of Nunu's good friends had also been executed. But her face quickly brightened.

Without her indefatigable will to live, she would have never lived to be 94. After the purge had ended, Nunu was one of the Tbilisi institute researchers who travelled to Red Army positions during the Second World War to fight disease: 'Once I had to go to the northern Caucasus because brucellosis had broken out among the troops.' This chronic bacterial infection had been brought into Georgia several years earlier by a race of sheep imported from Central Asia. 'I was supposed to find the source of infection in the mountains. The transmission of bacteria often took place through milk. Then I got infected myself. There wasn't any treatment for brucellosis back then. For ten years, I had aching joints and frequent attacks of fever, but I kept working the entire time.'

In 1945, when the war had just ended, the director of the institute dismissed her because of her illness. But even that couldn't break Nunu. No, not Nunu, who was always laughing during our meeting and said that she didn't mind walking down the six flights of stairs from her flat to go shopping when the lift refused to work. 'The lift is often out of order', she said, 'but now I'm getting off the subject. Let's go back to those dark times. When his friends disappeared, Eliava helped their relatives. For instance, he hired the daughter of construction engineer Jikia as a librarian for the institute. That was definitely dangerous.'

In September 1936, Stalin appointed Nikolai Ivanovich Yezhov as the head of the central secret police NKVD. His dwarfism, sickly constitution and cruelty soon earned him the nickname 'the bloody dwarf'.[55] In the annals of terror, his witch hunt is referred to as 'Yezhovshina'. No one was safe from Stalin's executor, who was said to have shot his predecessor himself. At the end of the bloody hunt, between 98 and 110 of the 139 former members and candidates of the highest Party committee were dead, although the exact number remains unknown.[56] Yezhov himself was eliminated in 1939. His successor was Lavrenti Beria. Eliava's two other powerful acquaintances, the veterans Mdivani and Ordzhonikidze, could offer no protection from Beria, since they were under pressure themselves.

In January 1937, Eliava travelled to Moscow with his wife, since she was negotiating an engagement with the Bolshoi Theatre. Their daughter Hanna later recalled that when her parents came home, they talked throughout the night. They seemed to be very worried. One by one, friends and acquaintances of the family were arrested. The house that had once been full of friends became a quiet place, with no visitors. On January 14, Nunu was invited to visit some friends. Eliava was there too. 'He sat there very quiet and alone. It was obvious that something was making his heart very heavy. Perhaps someone had warned him.' He escorted Nunu home, silent and sad.

She never saw him again.

That same evening, as Hanna and her parents were having dinner, soldiers forced their way into the flat and carried Georgiy Eliava off. An hour later, they returned and got his wife. Only hours earlier, the family had admired the construction work going on at the institute. Now both her parents had disappeared. Hanna remained behind, alone. She was 23 years old.[57]

Hanna never heard from her stepfather again. The henchmen then confiscated nearly all his papers, documents and photos. Georgiy Eliava, the charming pioneer of phage medicine, was not only dead, but for a long time all knowledge of him was erased from the memory of his fellow Georgians and the world.

Six months later, reports appeared in the newspaper about the trial that was to have taken place on 9 July 1937. Eliava was said to be a fascist, a traitor and a spy. He was accused of planning terrorist acts against the Communist Party, along with other conspirators, as well as contaminating the drinking water.[58] Contaminating wells was one of the standard charges in Stalinist trials, as was espionage or activities against the Party.[59] Eliava was sentenced to death. His stepdaughter thought he was shot the day he was arrested.

Eliava's powerful friends met a similar fate. Ordzhonikidze probably committed suicide after his deputy was sentenced to death and he saw no way out, and Mdivani was executed in summer 1937.[60]

On 3 April 1937, Eliava's stepdaughter Hanna was arrested and locked in a cell already housing a woman with white hair and a wrinkled face. It was her mother. They talked all night. When the guards realized that they had inadvertently brought mother and daughter together, they moved Hanna to another cell. Only a fat, screaming female guard appeared to have a soft heart. When she was on duty, mother and daughter met each other at the toilets. Soon Hanna was moved and she never saw her mother again.

In 1946, nine years after her arrest, Hanna returned from a camp in Kazakhstan. The state didn't officially announce the arrest of her parents until 1956. For Hanna, the only thing that remained was a few pictures from the family album.[61] The most charming surviving picture of Georgiy Eliava is in Nunu Kilasonidze's collection. It shows him surrounded by a bevy of young women, among them Nunu. They are all laughing. He looks pensive, though, as if he realized something would happen in the future.

**Figure 5.3** Georgiy Eliava surrounded by women (Nunu Kilasonidze, centre row, right of Eliava, and Elena Makashvili, centre row, far left)

## Why?

The secrecy surrounding Eliava's arrest left a number of unanswered questions. The most urgent of these was and remains – why? Rumour soon had it that Beria had his eyes on Eliava's wife. Another rumour claimed that two men whom Eliava had vaccinated against typhoid fever had died of shock, which meant that he had to die. After the trial, a newspaper wrote that Eliava had distributed typhoid fever among the people, killing 50.[62]

The numerous speculations correctly reflect one thing: in Stalin's reign of terror, there was no one reason why Georgiy Eliava had to die, but rather an entire range of reasons. And each was grounds for execution: Eliava's closeness to condemned veterans Mdivani and Ordzhonikidze, his contact with scientists in other countries or his attitude of being a free-thinking scientist. During this period of random killing, even having the wrong wife called for capital punishment. Eliava didn't have a chance.

What the 'bloody dwarf' unleashed and his successor Beria continued constitutes one of the gloomiest chapters of history. Georgian George Tarkhan-Mouravi described it as follows:

Not only the old Bolsheviks and acquaintances of Beria and Stalin became the victims of this terrible period. Nobody could be sure of his future, or of the future of his family. Any person could be arrested and executed within several days ... The NKVD groups made arrests late at night, so that the whole terrorized country would listen with painful attention to any late steps sounding in the night – one more victim is to disappear forever. Or, a sudden rumour would spread that someone was under suspicion, and even the closest friends would try to avoid meeting him ... A neighbour would easily inform the NKVD if there was a chance of improving the living conditions at the expense of the arrested ... Any absurd accusation could work, and torturers would make the victim accept his guilt whatever it could be. Perhaps to characterize the psychological atmosphere of the time it is worth recalling the situations when Stalin's image appeared on the cinema screen or he was named aloud. Nobody would dare to abstain from applauding, and such applause would continue for a very long time, with some people even fainting from the strain, as everybody was afraid to stop first – even this could serve as a pretext for the arrest.[63]

The reign of terror set quotas for arrested Trotskyites, spies and saboteurs that each district was to deliver. On a single day, 12 December 1937, Stalin confirmed the death sentences of 3167 prisoners.[64]

## The enchanted French villa in Tbilisi

D'Herelle and his wife could thank their lucky stars that their visas had not been renewed in autumn 1935. As foreigners, they would not have survived the bloody storm. In addition to their sponsor Ordzhonikidze, Kaminskii, the minister of health who had offered d'Herelle a job in Moscow, died as a result of the terror.

D'Herelle was hit hard by the death of his friend Eliava, says his grandson Claude-Hubert Mazure. His shock and disappointment are obvious in his memoirs. The chapter about his adventure in Georgia is only seven pages long, whereas hundreds of pages are devoted to other parts of his life.

In 1938, the house that the d'Herelles and the Eliavas were to have lived in was being built in Tbilisi. The Georgians had designed it in the style of a French villa in honour of d'Herelle. It still stands today, situated between trees in the park above the Mtkvari River.

Instead of the two families, the KGB moved into the house.

## In barracks and kindergartens

Stalin's executors may have had Eliava killed, but they couldn't extinguish his legacy. One year later, in 1938, the main building of the institute was finished. Its elegance and size made clear the hope raised by phage therapy. The plans for the phage palace harked back to Eliava and d'Herelle.[65] The model for some areas of the institute appears to have been the Pasteur Institute in Paris, where the two met. 'The labs in Tbilisi look astonishingly similar to those of the old Pasteur Institute', says phage researcher Hans-Wolfgang Ackermann, who worked at the Pasteur Institute in the 1950s and visited the institute in Tbilisi in 1997.

**Figure 5.4** 1950s photo of the entrance to the main building of the Tbilisi research institute for phages

In the numerous laboratories, Eliava's disciples who had escaped the terror did their research. Elena Makashvili completely devoted herself to Eliava's legacy. Nunu Kilasonidze claims that Makashvili was said to be in love with Eliava.

D'Herelle's assistant, Irakli Georgadze, remained loyal to the institute and later became its director. And Teimuraz Chanishvili, his successor, began his lifelong affair with phages a while later.

**Figure 5.5** Elena Makashvili (third from left), Georgiy Eliava's former assistant, at the Tbilisi institute in the early 1960s

The grandiose plans for a world centre with marble-clad specialized hospitals never materialized, however. Eliava, the initiator of the plan, had been executed as an enemy of the people. In addition, as the prospect of war increased, prestigious projects were out of the question. In future years, the institute produced a whole series of serums, vaccines and phages, especially for the military.[66] Like all armies, the Soviet army was afraid of dysentery, which could render entire battalions unfit for battle.

As early as the 1930s, the Soviets had tested phage therapy and prevention for dysentery in experiments, some of which were huge. This was more or less the same period that large-scale experiments on cholera took place in Assam, India. It was not Eliava and his team that led the way in this fight, however, since they focused more on typhoid fever and staphylococci.[67] Instead, the leaders at the front were institutes in the Ukraine and, later, in Moscow. V. A. Krestovnikova, a scientist from the Moscow Mechnikov Institute for Epidemiology, Microbiology and Infectious Diseases, provided exciting insights into this pioneer phase in a scientific review article published in 1947. During the war, Krestovnikova was an important phage researcher. According to her report, the Soviet scientists weren't as rash as their Western counterparts, who conducted their experiments with gusto. The Soviets designed their experiments systematically. Krestovnikova's portrayal of the situation needs to be taken with a pinch of salt, though, because it isn't clear how freely she was allowed to express herself. In addition, she left out a great number of details, although this isn't unusual for an overview article even today.[68]

The Soviets first investigated how phages behave in a healthy animal or human being. They concluded that the viruses were quickly excreted by healthy animals and if they were stored at all, then in the spleen, the body's rubbish sorter. This is in line with modern experiments.[69] Next, the researchers experimented with infected guinea pigs. Krestovnikova herself injected the animals with bacteria and then with phages in their abdominal cavities and observed what happened in the intestines: the phages indeed dissolved the bacteria inside the animal and multiplied as they did so. This was an experimental achievement that wasn't duplicated in the West until several years later, after 1942, when brilliant scholars like Dubos and Morton investigated how dysentery phages multiplied in infected mice (see Chapter 4).

This set the Soviet research apparatus in motion. By 1947 it had undertaken no fewer than 74 phage studies on dysentery.

All the experiments included control groups that were not given phages for the purpose of comparison, as Krestovnikova pointed out. This was another plus compared to the West. The phages were particularly effective in adults. When it came to the recovery rate of children, the Soviet researchers were not as satisfied. They focused instead on the preventive nature of phages in children. Prevention was at the top of the Soviet doctors' agenda, because Lenin had stressed the value of prophylaxis.[70]

In line with this, a team of 'the Soviet Union's most important epidemiologists' carried out experiments on prevention. 'Tens of thousands' of people were given phages as part of experiments in Moscow, Charkow, Kiev and Sverdlovsk. Once again, Krestovnikova's report didn't dwell on the details but focused on the results that indicated the effectiveness of bacteriophages. In the Ukraine, preparations were used that only contained the Shiga strain of dysentery bacteria, and the studies showed that there was only a preventive effect when Shiga dysentery was on the loose, and not if it was Sonne or Flexner's dysentery bacteria. Conversely, the Moscow-based Mechnikov Institute's remedy, which didn't contain Sonne phages, only protected the Leningrad subjects if no Sonne bacteria were going around. The preventive phage remedy was only effective if the phages were taken frequently and in large doses. This reflected the animal experiments that had shown that phages do not last long in a healthy organism because there isn't enough food to keep them alive.

At their meeting in Moscow in 1939, infectious disease specialists had officially acknowledged that phages were effective drugs in the fight against dysentery epidemics.[71] This gave rise to large-scale experiments that made the 1930s field trials in India seem tiny in comparison. Every year over a million children in daycare centres and nursery schools were regularly given dysentery phages during the diarrhoea season from May to October. The healing viruses made great strides during this period. The phages' new empire stretched from Leningrad to Vladivostok and from Tbilisi to the icy coasts of Siberia. Soviet phage researchers announced great successes with their dysentery prevention campaign using phages. After the national launch of the programme in 1943, one-third fewer children up to the age of three contracted dysentery than the year before. The actual role of phages in this success can no longer be determined, but at the time the prevention programmes were given credit for the positive outcome.[72]

## Rotting flesh and a sickly sweet smell

A powerful drug for dysentery was a blessing for both the normal population and the army. A second large-scale project was unmistakeably oriented towards military use. Its ambitious objective was to develop a remedy for gas gangrene, 'this most terrible complication of wounds inflicted in battle', as P. M. Zhuravlev, major general of the Soviet medical services, called it.[73] And indeed, gas gangrene was a grisly threat to any wounded soldier.

In the course of this infection, *Clostridium* bacteria settle deep in the injured person's wounds. They are among the ugliest kinds of bacteria, producing powerful toxins and avoiding oxygen. They are rotting bugs. Their toxins trickle into tissue, gnaw away at the cell walls, digest the connective tissue and liquefy the flesh. This destructive damage is accompanied by the production of gases, which collect in the dying tissue and bloat it – thus the name 'gas gangrene'. The Greek word for gangrene means 'an eating sore'. When the doctor at the front palpated the affected areas, he heard a characteristic crackling sound. The face of the restless patient turned yellowish and pale, while the soft areas surrounding the wound changed from brownish-blue to black and emitted a stale, sickly sweet odour. Fully conscious, the wounded soldier watched his own limbs decay before his eyes. The infection usually ended in death.

As Major General Zhuravlev wrote in *Microbiology and Epidemiology – Achievements of Soviet Medicine in the Patriotic War* published in 1943, Moscow-based researcher S. P. Zaeva isolated bacteriophages that were highly active against several types of clostridia and tested them successfully in animal experiments. With the attack on Finland in 1939, Stalin provided scientists with the opportunity to test them. A number of surgeons treated wounded soldiers using a cocktail consisting of gangrene phages manufactured by the Moscow Mechnikov Institute or the institute in Tbilisi. The leader in the testing was Georgian surgeon Aleksandr Petrovich Tsulukidze, a disciple of the murdered Eliava. He was especially suited for this endeavour, since he had already served as chief physician on the southwestern front during the First World War.[74] Of 767 soldiers who received phage medicine, 18.8 per cent died, while 42.2 per cent of a control group died who were treated with conventional methods – excision of the dying tissue. Mobile medical corps brigades that used phages directly at the Finnish front also managed to reduce the number of gangrene fatalities by one-third. A total of 10,000 soldiers participated in the study (probably without their explicit consent).[75]

On 22 June 1941, Hitler's Germany attacked the Soviet Union. Two teams of surgeons were sent to the front to brief military doctors on the use of phage mixtures on site and collect additional data on their efficacy. Doctors were instructed to excise the damaged flesh surrounding the wound and then inject 100 ml of phage mixture into the tissue. The huge amount of injection fluid made this procedure extremely painful. The surgeons recommended carrying out the injection while patients were under general anaesthesia or by adding novocaine to the phage mixture.[76]

## A golden era in misery

Whether the painful treatment was actually always carried out under anaesthesia is doubtful, however. The conditions at the front were appalling. Germany's attack caught the USSR off guard, since the two countries had signed a non-aggression pact. The Red Army was quickly pushed back. Countless research institutes and production facilities for drugs in cities such as Kiev, which were overrun by the German army, had to move to the countryside. Other research institutions suffered widespread damage from bombs. Hospitals were affected as well and were completely overcrowded.

The German army came dangerously close to the phage institute in Tbilisi. From the north, it moved through the Ukraine and around the Black Sea towards the Caucasus. Its goal was the oil-rich regions around Grosny and Baku. On 21 August 1942, the German Reich's battle flag flew high up on the Elbrus, at 5633 m, the highest mountain of the Caucasus. While Hitler's troops never reached Tbilisi itself, the Georgian people suffered a great deal during the war. An estimated 600,000 Georgians fought on the battlefields, and 300,000 died – one-tenth of the population. Throughout the Soviet Union, 21 million people died in the battle against Hitler's troops, 7 million of whom were civilians, and over 70,000 cities and villages were destroyed.[77]

It was a golden era for phages. While certain amounts of penicillin made their way to the Western Allies, starting in 1942, in the USSR the supply was extremely tight until the end of the war and beyond and at times was not available at all.[78]

Instead, the Soviets relied on phages in addition to sulphonamides. They made huge efforts to establish mass production to cover the army's needs.

At the front, phages came as a solution in vials, although often they didn't make it. The glass packaging was too fragile under these conditions. So researchers frantically tried to press phages into tablets. It took a while, but eventually F. E. Sergienko of the Central Institute for Epidemiology and Microbiology near Moscow came up with an answer. Instead of growing the phage-bacteria mixture in the usual complicated liquid culture, he grew it on plates of a thickened mixture of broth and agar-agar, a jelly-like product made of algae. Afterwards he treated the agar plates with chloroform vapours, which killed off the surviving bacteria but did not damage the phages, then scraped the agar mixture off the plate, mixed it with starch and equine serum, and dried and pressed it – producing a tablet.

Following this breakthrough, in a short time the Alma-Ata branch of the Central Institute produced 4 million tablets, the equivalent of 40,000 l of liquid culture. Quickly, researchers set up tablet production in several locations: Moscow, Sverdlovsk in the Urals and the Central Asian cities of Stalinabad and Tashkent.[79] In Alma-Ata, Kazakhstan, they tested the preventive capacities of dry phages on over 25,000 people. The tablets are said to have reduced the occurrence of dysentery to one-eighth of the previous level.[80]

There is no doubt that the war saw the height of phage production. While in Germany, dysentery phages disappeared from the army towards the end of the war, in the USSR, production rose by over 300 per cent from 1940 to 1942, with the increase in manufacture of other phage preparations jumping by 560 per cent.[81] In the above-mentioned book, subtitled *Achievements of Soviet Medicine in the Patriotic War*, 3 of the 15 chapters were dedicated to phages. Stuart Mudd, an American bacteriologist who visited the Soviet Union in autumn 1946, reported that during the war phages had been distributed in the soil and sewage and had been added to food.[82] 'The Soviet Union', noted phage researcher Krestov-nikova in her report of 1947, 'has truly become a second home for bacteriophages.'

## From blockbuster to clever niche player

After the war and the terror that didn't subside until Stalin's death on 5 March 1953, the Soviet Union had a lot of catching up to do. Foreign experts declared that medicine in the world's largest country was five years behind the West. Yet the USSR started to move forward. After the war, Soviet universities produced four times more doctors than their counterparts in the US. There was now even an institute for hygiene in the farthest corner of Siberia, and a network of collection centres for donations of mother's milk was put in place to guarantee that babies were well nourished. The USSR triumphed in other areas as well. On 4 October 1957, the beeping Sputnik 1 announced to stunned Americans that the Soviets were leaders in the space race. However, progress couldn't be measured across the board. US journalist John Gunther, who visited the country in 1956, described the paradoxes in the race to catch up. For instance, while Soviet ENT doctors had instruments so sensitive they could record a signal emitted by a single nerve, the lights in their operating rooms were completely antiquated.[83]

Soviet doctors also pushed the production of antibiotics, which were making the headlines in the West. Soon they were prescribing the easily administered chemical preparations, demoting phages to second place in the process.[84] Several institutes such as the ones in Tbilisi and Gorki[85] continued mass production of phages. They were primarily used for disease prevention in daycare centres and schools, but could also be found in hospitals and pharmacies. The largest purchaser continued to be the army, which bought 80 per cent of the preparations produced in Tbilisi. The institute's importance to the military was indicated by the fact that central government placed Eliava's successors more closely under its control. The institute was no longer supervised by the Georgian ministry of health, but was directly controlled by the health authority of the USSR.[86] The army set up a rigorous security regime. The fenced-in complex could only be entered after a security check. 'Although the security guards knew each and every one of us, there was absolutely no way to be admitted if you forgot your ID', says Inga Georgadze, who worked at the institute for years.

How is it that phage therapy continued to exist to this extent in the Soviet Union despite the availability of antibiotics? The response of critical doctors in the West has a heretical ring to it: phages had to work because the Party wanted them to. Critics believe that Soviet studies carried out during the war and post-war era do not meet today's standards – which is true. In line with this, they conclude that the efficacy of phage therapy observed by researchers at the time is the result of a sort of state doctrine.

From the point of view of today's scientific standards, it's difficult to judge how well the method served its purpose at the time. However, there is much evidence that contradicts the theory of mass delusion. Studies carried out during this period still appear to be better than those produced by most Western phage enthusiasts, if Soviet reports accurately describe the experiments. In addition, scientists like Zinaida Yermoleva studied both phages and antibiotics and were thus able to compare the two. And Yermoleva wasn't your average researcher, but the scientist who introduced penicillin and streptomycin in the Soviet Union.[87] The fact that the launch of antibiotics did not completely replace phages is a further indication that they were effective.

Production also probably functioned better in the East than in the West. For various applications, the Soviet researchers first crafted an optimal production method, as they did in the case of phage tablets, and then distributed this to manufacturers. This meant that there was a certain guarantee that manufacturers wouldn't fiddle about with unsuitable methods on their own. In addition, industrial production was monitored early on by a central location, the Priselkov control institute.[88] This centralized planning on the part of the Soviet system came close to d'Herelle's vision. All these factors probably led to better results than those achieved in the West and helped phages to retain their place in the drug inventory of Soviet physicians.

In some practices and hospitals, phages not only held their own, but were greatly esteemed. Soon the advantages of using phages instead of antibiotics became apparent. For example, if a paediatrician uses antibiotics for a case of inner ear infection, they not only destroy the 'guilty' streptococci in the child's ear, but also arbitrarily mow down other bugs – with negative consequences. The good bugs, millions and millions of which populate our bodies, protect us from being invaded by disease-causing bugs. If the good bugs are decimated by the random use of chemicals, pathogenic newcomers have an opportunity to step into the breach and trigger infections.

In the clean-swept intestinal tract of a patient, opportunists like *Clostridium difficile* can lodge themselves and lead to diarrhoea, and in seriously ill patients this can lead to conditions as serious as bowel perforation, sepsis and even death.[89] With certain antibiotics, up to 25 per cent of all patients treated suffer from intestinal infections with *Clostridium* that in turn need to be treated with antibiotics themselves. In one-fourth of cases, there are recurrences of the infection, and some hospitals even experience periods in which *Clostridium* epidemics rage.[90] The specific phages, on the other hand, only attack the harmful bacteria, leaving the protective microbe population on our bodies alone, and, in turn, do not lead to resistance. Phages are an intelligent drug.

Several antibiotics have even more harmful side effects. About 8 per cent of all people have allergic reactions to peni-cillin, and in about 1 in 200 patients the reactions can be extremely strong, even resulting in death. The group of amino-glycoside antibiotics is toxic for the kidneys and can damage hearing, and vancomycin can harm the kidneys. Chloram-phenicol damages bone marrow and can trigger conditions such as anaemia, which in rare cases leads to death.[91]

Some antibiotics used in the 1950s and 60s were much more harmful than the versions used today, which is why phages had loyal fans among Soviet physicians. Doctors in France, Switzerland and Germany, who continued to admin-ister phages that could be purchased from the Swiss company Saphal and the French Laboratoire du Bactériophage (see Chapter 4), seem to have based their choice primarily on the fact that they had fewer side effects, making them a sort of precursor generation to today's doctors who practise alterna-tive medicine. Several Soviet doctors even used phages to alleviate the damage done by antibiotics. In a hospital in Sverdlovsk, A. M. Litvinova used a mixture of intestinal phages and bifido bacteria to treat underweight newborns who had received antibiotics for pneumonia or blood poisoning. The phages were supposed to destroy the diarrhoea bugs plaguing the babies, and the bifido bacteria were intended to restore healthy intestinal flora. A similar remedy was used for cancer patients.[92] Phage therapy led a secure and quiet existence in the bosom of Soviet medicine.

## Rescue from the sewer

In the early 1970s, trouble was brewing in the Soviet hospitals. Infectious disease specialist David Shrayer, who later emigrated to the US, was still living in the USSR. He remembers every detail of an emergency operation along the enormous construction site of the Baikal Amurskaya (BAM) railway, which was to connect the Pacific coast with the western part of the country on a 6000-km line. Staph infections were rampant among the workers, who suffered from pus-filled wounds, abscesses and furuncles. 'It was an alarming epidemic', Shrayer says. 'At some of the construction sites, despite the Siberian cold, the workers lived in tents in which the temperature never made it above freezing on icy days.' The unpaved roads, that the frequent rain turned into mud, were full of dogs, rats and rubbish.

In some hospitals along the railway line, infected patients and doctors walked unchecked from department to department. Often 10–12 patients with open infections were crammed into one room. In one bacteriological lab Shrayer visited, sewage dripped from the ceiling. The staphylococci ran amok, spreading quickly in the hospitals. Many people were infected by contaminated food and stray dogs. The bacteria were highly pathogenic and multi-resistant.[93]

Conditions were particularly bad along the BAM railway line, but they were unbearable in other places as well. Too often, poor hygiene and negligence caused doctors to prescribe antibiotics at the drop of a hat. This resulted in a fateful spiral, with more antibiotics giving rise to more resistances that were treated with still more antibiotics leading to still more resist-ances. It wasn't long before antibiotics were ineffective. A wave of resistant microbes swept over the country.

The medical authorities hoped that phages would save the day. Teimuraz Chanishvili, deputy director at the Tbilisi insti-tute at the time, says: 'In 1975, a decree was published. All the country's epidemiological departments had to send their bacteria samples to Tbilisi so that we could find new phages that were active against the resistant microbes. I wrote the decree myself and brought it to Moscow to be signed.' Soon countless samples arrived from all over the Soviet Union – 20,000–30,000 every year. 'It was an enormous amount of work to diagnose, test and store all the samples. In many of them there were several different germs that we first had to separate using a complicated procedure.'

Chanishvili and his team were ready for a renaissance: 'A few years earlier we had decided to improve our phages. In the meantime, we knew more about the viruses than our predeces-sors had and we wanted to use this knowledge.' This gave them the idea of selecting new dysentery phages that were particularly suited to the location they were to be used in – the intestinal tract. To do this, they treated different dysentery patients with various phages and selected the phage that had remained longest in the intestines to be used in future treatment.[94]

On closer inspection, the scientists at the Tbilisi institute real-ized that a preparation developed to treat staph in Tbilisi could be inactive in Moscow. The high specificity of the phages, one of their best features, turned out to be a big problem for centralized production. They also discovered that they could never take a break. Keeping up with the changeable bugs condemned them to non-stop research. The bacteria stub-bornly defended themselves against attacks and became resistant to antibiotics or phages.

Bacteria and phages have been fighting a constant battle from the start of their existence, probably for over three billion years. A bacterium that is plagued by phages modifies a surface protein on which a certain type of phage docks and, from then on, the parasite has no way in – unless some phages contain a suitable mutation that keeps them in the game. This endless ritual of evolution creates a type of 'arms race' between the two opponents.

Recently, researchers from the US have revealed the particularly crafty procedure of a phage. *Bordetella* bacteria, which cause whooping cough, switch back and forth from one stage of life to the other. One stage attacks the human victims, while the more peaceful form lives in the environment. The phage called BPP-1 normally only attacks *Bordetella* when the bacteria are set to strike humans. When the scientists had a closer look, however, they discovered several viruses that only attacked *Bordetella* in the environment and others that attacked both forms. The phages switch back and forth between the various forms by changing their tail fibre proteins, which they use to dock onto their victims.

Virologists soon realized, however, that the ingenuity of phages is much greater. They have their own apparatus that makes a certain amount of mistakes when switching the tail fibre proteins. It is true that some phages with mutated tail fibres no longer attach to bacteria, but to make up for this, variants are constantly emerging that counterattack the mutations of the bacteria. It is an ongoing battle in which phage BPP-1 will probably never be the loser. The phage can theoretically construct a dizzying 9200 trillion types of tail fibres in order to stay one step ahead in the race.[95]

Mother nature keeps coming up with new phages to keep resistant bacteria in check. Chanishvili only had to look for them, and he and his comrades wanted to do this systematically. They designed a clever system that could probably only function in the centralized USSR. Phages and Communism flourished well together. The basis for this success was the decree that Chanishvili himself had written. Now his scientists continuously tested bacteria that arrived from all over the Soviet Union to see whether they could be destroyed by the available phages. If they proved to be resistant, the researchers opened their fridges and looked for other phages in their growing collection to do the job. If they succeeded, they expanded the existing phage cocktail and checked this bacterium off their list.

However, the scientists often came up empty-handed. This meant that they had to pull on their boots and hunt for new phages. 'To do that, you need to have an imagination and military discipline', says Liana Gachechiladze, who joined the Tbilisi institute in the 1960s. 'Most of the phages are in places where they can find food: in a hospital where resistant bacteria have turned up. The easiest way to do it is to collect the phages in the hospital's sewage, since they all gather there.' An even simpler method is to go where all the city's waste water collects. In this case, it was the 'cloaca maxima', the brown Mtkvari River where Eliava had once found cholera phages. No one who has seen the Mtkvari River in Tbilisi has trouble understanding why boots are the footwear of choice for collecting phages.

Military discipline is required to carry out the next step, in which the right phages are isolated from the sewer. In one of the labs in the old Tbilisi institute, Lamara Chanishvili showed us how this works. Like Liana Gachechiladze, the 75-year-old scientist still works in the institute every day. She was an instructor for a long time and showed new generations of scientists how to get new phages. 'We take the water sample that we suspect contains phages, put it in broth and inoculate it with the bacterium for which we're looking for a remedy', Chanishvili explains. After a growth phase of 18–24 hours at 37 °C, she centrifuges the solution. The phages, which are light, remain in the solution, while the heavy bacteria are pushed to the bottom of the centrifuge tube. Then Chanishvili uses a glass pipette to suck up the phage solution and filters it in order to remove the last bacteria. The phages remain in the filtrate. 'Then we spread the filtrate on agar plates on which we have seeded the matching bacteria. We put the plates in a warm place again overnight so that the bacteria will grow and the phages can multiply on them.' The result is the characteristic holes in the bacteria lawn, referred to as 'plaques'. Chanishvili uses sterile wire eyelets to grab the phages out of a plaque and spreads them on a plate with the same bacteria. New plaques emerge. She repeats this washing procedure five or six times until the shape of the holes remains the same, an indication that she is now dealing with only one type of phage. This finding is confirmed by looking through an electron microscope.

Military discipline is also required in the final act. Now the new phages have to prove that they have the power to dissolve the entire horde of stubborn bacteria. To do this, the phages are spread on an agar plate with each bacterium to be tested in order to see whether holes form or not. Chanishvili or her pupils often herded several hundred microbial strains through the long, drawn-out series of tests. And all this took place in labs that have remained mostly unchanged since Eliava planned them, using equipment that he ordered himself: Reichert brand microscopes, imported from Germany in the 1930s, and wooden incubators that Eliava himself may have stood in front of, happily waiting to remove his phage cultures.

The scientists added the phages that survived the selection process to the existing mixtures, and removed phages that had become ineffective. They repeated this complicated procedure every six months for each phage drug in order to keep up with the agile bacteria. As a result, pyophage, the mixture for purulent wound infections, contained phages active against *Staphylococcus, Streptococcus, E. coli, Proteus and Pseudomonas aeruginosa* in a constantly changing composition. Different types of bacteriophages were added to fight each type of bacteria in order to make allowances for the picky appetite of the viruses.

The combination strategy also had the advantage that it kept bacteria away from a number of paths to resistance. Bacteria become resistant to phages when, for example, a mutation changes the location where a phage attaches itself. If there are two types of phage in one preparation that attack bacteria at different locations, only those bacteria are resistant that are mutated at both locations. The probability of this type of double mutation is much lower than that of a single mutation, which means that it happens much less frequently. If there are more than two different types of phage contained in the mixture, the probability of resistant bacteria occurring drops even more.

**Figure 5.6** Searching for new phages at the Tbilisi institute in the early 1970s

The emergency service that Tbilisi institute researchers used to counter the creative power of bacteria was sanctioned and regulated by Soviet health authorities. The preparations available for purchase were required to destroy at least 70 per cent of a determined set of test bacteria. They were also regularly tested for their toxicity in animal experiments. 'By using these measures, we could drastically increase the rate of success', says Teimuraz Chanishvili. 'It rose from an average of 57 per cent in the 1950s to 90–95 per cent in the 1980s.'

This vast experience in detecting and selecting new phages that Chanishvili and his colleagues garnered over the decades is viewed with awe by some Western scientists. 'Their know-how is unrivalled', says Sergey Bujanover of the Israeli company Phage Biotech. The ability to quickly gather new phages against lethal bacteria from sewage was and remains the core of the Eastern variety of phage therapy. If some researchers had their way, it could also serve as the basis for a renaissance in the West.

## Heroic deeds

In France, a similar expertise is available – but hardly anyone knows it. One of the few French researchers who still knows his way around phage therapy is 80-year-old Jean-François Vieu. In 1956, the physician joined the Service du Bactériophage of the Pasteur Institute in Paris, some 40 years after d'Herelle had started his phage experiments there. The task of the Service du Bactériophage was not phage therapy, however, but rather basic research. Using methods similar to those of the Soviet researchers, Vieu established a sort of fire brigade for infections: 'We only got involved in emergencies, like when a doctor called and said he had a patient that no antibiotic would help', says Vieu.

There were cases like this even then, when Vieu maintained his free emergency service from 1956 until the 1980s. He calls it the 'heroic phase' of phage therapy. It was not a period of elaborate studies. Since Vieu only treated individual patients, no one could scientifically prove whether the phages had brought about recovery or whether healing came about spontaneously. All the parties involved were simply happy when the patient could be discharged in a recovered state.

Vieu remembers one case particularly well. It happened back in 1962 or 1963 when he was at a conference in Mexico City. A Mexican doctor told him about a 20-year-old patient who had had a laparoscopy and then contracted an infection at the incision where the endoscope had been inserted. The doctor had been treating the large boils around the incision for six months without success. Vieu offered to help him, although he didn't have much time, since the conference only lasted for 12 days.

'The first thing I did was to take a sample from the purulent area in order to determine the type of bacteria', Vieu relates. Instead of attending the conference, he spent the next 12 hours in a lab to find out quickly what kind of bacteria he was dealing with. It turned out to be a whole collection: *Enterococcus, S. aureus, E. coli, Proteus vulgaris* and *Providencia rettgeri*. It isn't unusual to find several bugs at a single site of infection.

Since Vieu hadn't brought any phages with him to Mexico, he used the Soviet researchers' method and went on a search for active phages in waste water. After working for four days and nights, he had gathered four active phages, none of which was effective against the *Providencia* bacteria, however. Despite this, he injected the combination of phages into the patient's boils. 'I poked the needle of the syringe into the infected site and slowly pulled it out while I pushed the solution through the needle', Vieu says. 'By doing this, I was able to distribute the phages in the entire furuncle, a method that goes back to d'Herelle.' Within a few days, the four types of bacteria disappeared from the patient's boils and only the *Providencia* bacteria stayed behind, as expected. 'The Mexican doctors were dumbfounded.'

In order to bring about a complete recovery, Vieu smuggled a sample of the *Providencia* bacteria through customs on his way home. Back at the Pasteur Institute in Paris, he isolated a matching phage that the Mexican doctors were able to use to completely heal the furuncles. 'Afterwards, the patient wrote to me that she prayed for me to the Holy Madonna.'

Vieu's services were in great demand. Every year he received between 40 and 120 appeals for help. 'A number of infectious disease specialists told me that phages were their ultimate remedy.' Vieu wasn't the only scientist in France to offer his 'heroic' services. There were also researchers in Strasbourg and Lyon who practised phage therapy in emergencies. Like Vieu, they are all retired now and have been forgotten, along with their know-how.[96]

## Chanishvili's quest for the Holy Grail

In the Soviet Union, despite the phage researchers' offensive in the 1970s, several extraordinarily insidious, resistant staph could not be conquered. They lurked in hospitals and installed themselves into a number of patients on a chronic basis. 'It was a catastrophe', says Teimuraz Chanishvili. 'A special meeting was called in Moscow to address the state of emergency. I recall a particularly dramatic talk. The speaker said that the situation was so bad that pregnant women had to be advised to have their babies at home again.' In this predicament, an idea came to Chanishvili that no one had dared to think about in the Soviet Union for 30 years: what was needed were phages that could be injected directly into the bloodstream to destroy the marauding bacteria there. That would be a powerful remedy for counteracting the increasing number of chronic infections.

Like their colleagues in the West, Soviet doctors had occasionally injected phages into the bloodstream of their patients until the Second World War. The violent reactions in some cases, however, which included attacks of fever and chills, caused them to abandon this treatment method. The preparations were simply too impure, because the manufacturing methods at the time left behind a wild mixture of phages, bacterial debris and broth proteins. This triggered a momentary and dangerous overreaction on the part of the patient's immune system, causing it to be more like shock therapy than phage therapy. Starting in the 1950s, Soviet doctors began dispensing phages primarily as tablets and solutions for oral administration or for irrigating wounds.

However, a phage that could be administered intravenously remained the scientists' Holy Grail and Chanishvili began his search again: 'I spent 15 years of my life working to find a solution.' Despite the long years of searching, even the cataracts in the eyes of the elderly scientist cannot hide the satisfaction he radiates when he talks about his quest. The first goal was to manufacture phages in a way that left nothing available to the immune system that would cause it to run amok. To do this, Chanishvili developed a broth that was manufactured synthetically instead of from beef. It took him a huge number of trials to find a way to cultivate phages in a high enough concentration. Countless numbers of rabbits were used in experiments in which Chanishvili convinced himself that his phage worked. There was no violent overreaction on the part of the immune system, but the staphylococci were completely removed from the rabbit blood – and it worked for 95 per cent of all multi-resistant staphylococcal strains. Healthy human subjects also tolerated the phages well.

However, the project suddenly came to a standstill. 'We had an enormous amount of trouble finding doctors who were prepared to test the preparation in patients', Chanishvili says. 'No one dared to do it. We were truly pioneers.' The goal was to find a patient who couldn't be helped with conventional remedies and for this reason was prepared to take the risk. And then a doctor had to be found who was willing to break the taboo. The patient turned out to be a young man named Avtandil Chkheidze. He was suffering from a chronic purulent skin infection that even antibiotics imported from the West couldn't cure. One after another, furuncles erupted from his skin. Chkheidze was miserable. He was on the verge of needing crutches and didn't worry about the risks associated with being a guinea pig. Professor Vakthang Bochorishvili, a renowned physician from the Sepsis Centre in Tbilisi, dared to carry out treatment. After three days of infusions with Chanishvili's phages, the boils had disappeared and five days into treatment Chkheidze went to a party with his friends.

Bochorishvili and Chanishvili initially expanded the experiments to include 20 patients and then treated a larger number of subjects. Several hospitals in Tbilisi and Moscow participated. A paediatrician on the intensive care ward for newborns used the new drug for infants for the first time. However, as a precautionary measure, the director of the hospital insisted that the phages first only be administered along with antibiotics. Of 98 babies who received only antibiotics, 8 died, while only 1 of the 148 babies who were injected with the combination of phages and antibiotics died. There were no side effects reported in any of the studies.[97] 'Only then did we start to mass produce intravenous phages', Chanishvili reports.

In the mid-1980s, phage therapy and the institute in Tbilisi flourished again. When Chanishvili uses the term 'mass production', he isn't exaggerating. The institute produced 80 million tablets per year for treating dysentery and typhoid fever alone.[98] Lamara Chanishvili (no relation) was once the head of a production unit for dysentery phages. She still remembers what it was like when the institute churned out bacteriophages the way a brewery produces beer – by the hectolitre. On production alone, there were 800 employees. On the first floor of the institute, in a sort of industrial kitchen for microbes, women turned huge amounts of beef and other ingredients into broth. It was pumped through steel pipes to the third floor, where the mighty fermenters stood in the tiled manufacturing halls. In every room, there were five gigantic steel machines with a stirrer, pressure gauges and a tank that swallowed 500 l of beef broth. Overnight, the cleaned incubator vats dozed in the sterilizing, toxic blue ultraviolet light and waited for the bacterial cargo that was germinating next to them in huge glass flasks. In the early morning, the produc-

**Figure 5.7** Mass production of phages at the Tbilisi institute in the 1950s

**Figure 5.8** Workers pump finished phage solution through the round sterile filters at the Tbilisi institute in the 1970s

tion crew filled the fermenters with broth and the starter cultures from the glass flasks. This meant five times 500 l, for a total of 2500 l. If demand rose, the women had to work two shifts. They wore white aprons, a bonnet and, depending on the production step, a surgical mask.

Later in the morning, they added the phages. If the phages had completed their task in the tanks after a few hours, destroying the bacteria and multiplying exponentially, the workers used powerful vacuum pumps to suck the broth through sterile filters the size of car tyres. Now the piecework began. The women had to put the 2500 l of phage solution into glass vials by hand. In completely sterile surroundings, they poured in 10 ml of solution from the tube, fused it shut and repeated the whole procedure again, 250,000 times per shift and production unit, day in and day out. One worker normally managed 500 to 600 vials, and the record holder filled 1000 of them on her shift. In 1970, a machine was finally produced to do the work. If the figures are extrapolated, one group operating five fermenters produced over 600,000 l of phages per year.

The army continued to be the bulk buyer of the solution. A number of hospitals throughout the country in cities such as Moscow, Leningrad and Kemerovo, however, were also using phages again. In Tolyatti, the car-manufacturing metropolis, doctors rarely used antibiotics, says Zemphira Alavidze, who joined the Tbilisi institute in 1968. The phage researchers' fire brigade service proved to be a success. If a new troublesome bug turned up in a hospital, the scientists swarmed out and searched for a solution in the sewer. This was the case when an epidemic of *Serratia marcescens* ran rampant in a Tbilisi children's hospital. This bug is a stubborn opportunistic bacterium that infects burn victims or other susceptible patients and is often resistant to antibiotics. *Serratia* bacteria had infected 350 newborns when Liana Gachechiladze was called in to assist. Her team promptly found phages in the Mtkvari River that were used to disinfect the newborn ward. After the phages were thoroughly tested, they were administered to the staff as a preventive measure.

In the meantime, the political climate had improved. After the horror of the Stalinist terror, the Second World War and the long, leaden period of the Cold War, veterans like Nunu Kilasonidze and Teimuraz Chanishvili watched Mikhail Gorbachev's glasnost and perestroika emerge. In addition to hundreds of other new freedoms, it was suddenly permitted to ask about the missing Georgiy Eliava. Throughout the years, phage researchers had kept their ancestor in their heads and hearts. In 1989, after repeated requests to the higher authorities, the Tbilisi institute was finally allowed to use the name of its founder. Since then, it has been called the Eliava Institute for Bacteriophages, Microbiology and Virology.

It seemed as though a golden era had finally started.

# 6

## keepers of the grail in peril

The machines were running full blast. Every day the kitchen brewed thousands of litres of broth. The pipes pumped the soup into the vats, which incubated the bacteriophages. Powerful filters cleaned the masses of viruses. Compressing machines formed the dried phage powder into myriads of tablets. On that day back in spring 1989, production head Amiran Meipariani delivered 88,600 pills for diarrhoea and 497,000 for salmonellosis prevention to Central Asia. It was a big delivery.[1] And it would be the last one.

At the height of research and production, everything came to a grinding halt. It was the beginning of an agonizing period. Ironically, it was Gorbachev's long overdue reforms that brought misery to the Eliava Institute. In Georgia, as in the rest of the Soviet Union, perestroika and glasnost inspired feelings of freedom. By late 1990, free elections had taken place and on 9 April 1991 the tiny republic declared its independence.

For the country, this triumph was deceptive, and for the Eliava Institute it was a catastrophe. When the Russians realized that Georgia wanted to secede from the country, they stopped payments, and demand for bacteriophages from the Red Army and several republics where dysentery was rampant collapsed. This was the beginning of a period of suffering for phage therapists that has yet to end.

At the beginning, the mighty production facilities were broken up and privatized and ties to the institute cut, the 'head' as it were. The researchers were left behind in their labs, without salaries or research money. Meanwhile, production dropped to a minimum. The new owners mostly focused on other enterprises, even selling off the equipment. Only one venture, called Biopharm, managed to pull together the resources to produce a variety of pharmaceuticals, including small batches of phage drugs for local pharmacies. But it did not have any scientists in its ranks to continually adapt the phage products – the process the formerly used production scheme was built on. There were constant power outages and at times even the water supply was cut off, forcing the owners of the small companies to go to the institute's park to drill for water themselves. Soon, many of the production halls were desolate and empty except for some rusty scrap iron and piles of mortar.

**Figure 6.1** A phage production hall at the Eliava Institute in the summer of 2002. Only one of the fermenters survived – shut down and rusty

The once flourishing institute decayed. The stables that had once housed 50 horses for serum production stood empty.

The proud library was boarded up. A mass exodus started. In the institute's heyday, some 1200 people worked here: about 250 were researchers and the rest worked in production. Today there are only 70 employees, many of them older women. The younger employees looked for other ways to make a living. Those who stayed behind had to find a way to stay afloat, since making ends meet on the $30-a-month salary provided by the Georgian government – when it managed this – was impossible. The institute had no money to pay for materials or wages.[2]

Inga Georgadze was the head of the institute's virology lab. In order to support herself and her department, the trained doctor opened a tiny diagnostic and medical practice called Diagnos 90. It was housed in the former security building at the entrance to the institute's park. Some years later, Alfred Gertler, the musician with the infection in his ankle, would be examined there (Chapter 1). Diagnos 90 consisted of three examining rooms, an office and a small pharmacy. An old picture from a calendar on the wall, a small table and a shelf hanging askew kept the place from looking completely bare. After Georgadze had set up the practice, many doctors from the city started to send their patients there to get an exact diagnosis of a particular infection. Georgadze offered advice on the best-suited medication – which often meant prescribing phages. When this was the case, she sold the patient one or two flimsy cardboard packets containing bacteriophages bred in the primitive, small-scale cultivation facility that one of the institute's research groups had improvised.

In order to earn this extra money, the researchers had to give up their original work. Liana Gachechiladze was 61 at the time of the break-up of the Soviet Union. Until then, she did a lot of basic research. Her old-fashioned flowered dresses might make a Western visitor mistake her for a retired grandmother. Yet

Gachechiladze was doing sophisticated studies to expose the enigmatic immunological and infection properties of bacterial viruses. After the collapse of the Soviet Union, however, she also had to rethink her options. 'I used to do exciting basic research, but since 1989 we have to make what sells.'

Tarasi 'Tato' Gabisonia, another loyal staff member, survived because of a seemingly impossible level of activity. As well as his position as head of the institute's microbiological lab, Gabisonia held lectures at the university for veterinary medicine and did consulting for Georgian biotech companies. If I wanted to chat with him about his work, I often ended up sitting with him in his rusty Lada, which was filled to bursting with his bulky figure. The summer heat and the huge suit that Gabisonia always wore made the sweat drip off his brow. Steering the car over the pothole-ridden streets of Tbilisi, he gesticulated wildly as he explained the advantages of phage therapy. The conversation ended when we reached one of his workplaces.

The first few years after Georgian independence were particularly difficult. In the winter, the temperature in the labs was a mere 5 °C. There were two or three hours of light a day. Meipariani, then 67, the head of production with nothing to produce, would sit in his tiny office behind two dead phones wearing a cap and scarf, chain-smoking to produce a little bit of heat. He could have chosen to retire, but then his income would have dropped from $30 a month to $8 in pension payments – with bread costing 85 cents a kilo. At least the valuable phage collections weren't threatened during the winter. 'We were frozen, so that meant that the phages were okay', says Nino Chanishvili, a researcher at the institute and niece of Teimuraz Chanishvili, the former director of the institute.

It got even worse. Georgia was overrun by chaos. The collapse of the Soviet Union rekindled the old ethnic conflicts, which nearly pushed the fragile new country over the edge.

First, the Ossetes, who live in the north, wanted to use violent means to fuse with the Ossetes in Russia. Then there was an uprising by the Abkhazians, a tiny ethnic minority. Wars broke out twice. Then came the civil war. Zviad Gamsakhurdia, the first freely elected president, turned out to be a despot who used brutality to deal with the ethnic minorities in Georgia and failed to control the growing economic chaos. Increasing opposition and unrest culminated in open battles.

In the centre of Tbilisi, there was a two-week-long show-down with tanks and machine guns that claimed the lives of at least 100 people. Georgadze recalls: 'Nothing worked. None of the buses were in operation, so some of the colleagues had to walk three hours to the institute every day. Despite this, we kept the labs and Diagnos 90 open.' Zemphira Alavidze even worked feverishly night and day to develop a phage spray for the ailing Georgian troops; it is claimed to have saved many lives during the Abkhazian war. The soldiers carried the phage spray with them. When an injury occurred, they had to spray the phages on the wound immediately – 3000 soldiers with penetration wounds and shattered limbs were treated this way on the battlefield. Only 12 of these fighters died. Amputations were not necessary. 'That was a shining result that turned battlefield surgery upside down', relates Nodar Daniela, a Georgian surgical veteran. 'In most cases, amputations have to be done because of the huge risk of infection.'

## A romance between West and East

In this way, researchers worked their way from one icy winter to the next and from one power outage to the next, and the institute decayed even more. It had been ages since the gaping entrance hall and corridors of the labyrinthine complex had had lights that worked, but that didn't particularly bother the stray dogs that sought refuge there. The wooden floor was riddled with craters, and plaster crumbled from the ceilings and walls. The gigantic centrifuges, relics from the Soviet era, broke down one after the other and machinery stood collecting dust. Then, in 1996, US journalist Peter Radetsky took a trip to the forgotten city of Tbilisi. He talked to Chanishvili, the elderly scientific director, and was amazed as he toured the ruins of a medicine that hardly anyone remembered in the West. In the US popular science magazine *Discover*, he revealed the 'good viruses' to the public that Georgians were using to kill bacteria.[3]

One reader was particularly taken by the report. Canadian financier and multimillionaire Caisey Harlingten was sitting on a plane with his friend Monica. As he thumbed through *Discover*, he became absorbed in the article about the history of the odd medicine being practised on the other side of the world. The venture capitalist was fascinated. A seemingly undiscovered opportunity was knocking. The West was teeming with antibiotic-resistant bacteria. People were dying, and these impoverished Georgians seemed to have a remedy for combating them. A deal just had to be in the offing.[4]

Harlingten didn't skip a beat. He contacted US phage researcher Elizabeth Kutter, who the article reported had connections to Georgia. In fact, it was Betty Kutter who had paved the way for *Discover* reporter Radetsky to go to Tbilisi. During a stay there in 1990, she came upon the institute in the Soviet Union, which was in the midst of breaking up. Three years later, she paid another visit there. Kutter is a professor at the Evergreen State College in Olympia, Washington, who has studied the molecular biology of phages since 1963. At the beginning she was rather sceptical about seeing her lab pets being used as therapeutic agents. But she gradually became a devotee of the Georgian phage magic. 'The more I saw what the researchers there were doing and the depth and breadth of what they had achieved, the more impressed I was.' She began supporting her friends there in any way she could.

Harlingten's inquiry appeared to be a promising opportunity. Kutter arranged a visit and just a few months after the publication of the *Discover* article accompanied him on a visit to Tbilisi. In nine months, the financier raised money and founded Georgia Research Inc (GRI). Harlingten wanted to buy the exclusive rights to all the know-how of the Eliava researchers at a rate of $75,000 a year. With the help of Nino Chanishvili, he set up a lab in a privatized part of the institute. Soon the particularly nasty microbes arrived that Harlingten's staff had collected from dying patients treated in hospitals all over the US. Chanishvili and her team scanned their large phage collection for viruses that would destroy these bacteria. Scientists at the US headquarters of GRI were supposed to develop and produce the life-saving phages as drugs in line with strict American standards. The institute in Tbilisi would earn a small commission for each approved preparation.[5]

Finally a breakthrough had come.

The BBC sent a TV crew to report on the Eliava Institute in the faraway Caucasus and its marriage to its Canadian 'sugar daddy'. Harlingten told the TV crew: 'The cash flow hasn't started yet. We're talking about a pharmaceutical development company. You know, it won't be long before it will be a matter of hundreds of millions of dollars. I'm absolutely sure about that.' Chanishvili, a woman of few words, joked hopefully: 'I'm old now, and phage research is my hobby. I hope it will make me wealthy.'

Bubbling with enthusiasm, Harlingten arranged a conference on phage therapy in a former seat of the Communist Party in the mountains outside Tbilisi. In addition to the Georgians, he invited some of the big names of phage research in the West. The Western scientists were supposed to use their expertise to help Harlingten size up the situation. They must have all felt like pioneers in the ex-Communist mountain enclave, where the toilets didn't work, the beds were broken and the lamps stayed unlit.[6]

## War of the Roses

The romance between West and East fell apart, however. Harlingten and his CEO Richard Honour began to doubt whether in the desolate surroundings exotic drugs could be developed that would survive the strict requirements posed by the US Food and Drug Administration (FDA) and appeal to American consumers themselves. 'Do you have a lethal infection? Then take viruses from the Eastern bloc to save your life!' It didn't sound like a convincing Madison Avenue pitch.

The loquacious Honour, who today is head of the American company Viridax, didn't beat about the bush: 'I shocked the people from Eliava when I held a lecture about the FDA and revealed that we could hardly succeed in Tbilisi because it would cost millions to set up suitable labs. They nearly threw bottles at me when they heard that. I had shattered their dream.' Soon after, Harlingten's bearer of bad tidings closed down the branch in Tbilisi. 'I couldn't wait to get out of there', says Honour. 'In the short time they spent here they took a lot of know-how with them', says Nino Chanishvili bitterly. Now, Viridax is working on its own variation of phage drugs that is more likely to get FDA approval. Caisey Harlingten refuses to respond to questions concerning the mismatch. His GRI became Phage Therapeutics. When it got into financial trouble, Harlingten left in 2000 to help found Regma Biotechnologies, which changed its name to PhageGen in 2003. However, recently, PhageGen has dramatically changed its business activity and is now focusing on the search for gold.

## Grounds for divorce

A look at the rudiments of the phage therapy system the Eliava researchers salvaged from the Soviet era makes clear the reservations that Honour and Harlingten must have had. One of them is the problem of the cocktails. Most phage medicines from Tbilisi consist of a mixture of phages rather than a single one, so that the primary strains and species of bacteria of an infection can be attacked. This was allowed in the Soviet Union, and it made sense. Mixtures of this type are poison for the hordes of microbes – as they probably are for the strict FDA. In the eyes of the FDA, there are too many unknown complex viruses involved. It sees a drug that consists of a bacteriophage as an exotic novelty, but an entire cocktail of wild phages is an impossible novelty. 'The FDA indicated that we could only start with a single phage', says Richard Carlton of the US phage company Exponential Biotherapies.

Even if a phage mixture managed to gain approval, there would be other problems. The Soviets continuously fine-tuned their cocktails to the changing microbes in hospitals. The people at the Eliava Institute continue to do this today. For the drug authorities in the US or Europe, this biological warfare is unfamiliar, which means it is unclear how they would react to it. The most similar process used in the West is probably the influenza vaccine, which is reformulated each year to match the circulating flu viruses. If the drug authorities were to require separate approval for each new mix, phages would be prohibitively expensive. Betty Kutter thinks that the authorities should differentiate between the topically applied phage preparations and ordinary pharmaceuticals: 'Today, phages are used for external applications in Georgia. This resembles natural products that are used extensively in many parts of the world', she explains. 'And these are often licensed in various ways for medical applications without complete chemical characterization.' She uses maggots as an example. The larvae of flies are used by some doctors in Europe and the US to clean chronic wounds infected with antibiotic-resistant bacteria. The maggots are placed on the wound, where they eat the bacteria and dissolve the diseased and dead flesh – but not the healthy tissue close by. Interestingly, maggots, like phages, were already used by doctors until the introduction of antibiotics, only to be rediscovered in the 1990s. 'If phages are to help solve our rapidly growing antibiotic crisis, alternative regulatory models need to be explored', Kutter says. 'Of course, they still have to ensure appropriate levels of safety.'

The medical obstacles that confronted the collaboration between Eliava researchers and US businessmen were no doubt compounded by cultural ones. This is something that neither side talks about, referring to 'culture shock' at best. In this context, it's interesting to hear Sergey Bujanover of the Israeli company PhageBiotech talk about his experience. Bujanover emigrated from Russia to Israel a long time ago. He is familiar with both worlds. His company made a deal with one of the three to four companies that have survived from the Soviet phage imperium, an institute in Ufa that now produces phages for the Russian market. PhageBiotech bought the 'unbelievably rich' phage and bacteria collection from Ufa – and took 'two years and ten times the purchase price in additional expenses until all the required signatures were acquired', says Bujanover. 'Twenty Western companies before us had tried to do the same. All of them returned empty-handed. It's hard to do business with people in the former Soviet Union. You have to understand them. They aren't able to think five years into the future. The thinking is that you have to survive today and that's enough. So they want to earn as much money as possible right then and there.'

## Instead of a share of a billion dollar market, a life in shambles

The first disappointing flirtation with money had left its mark on the Eliava Institute. The scars remain. Who profited from Harlingten the most? How good or bad was the deal really? The researchers in their crumbling labs caught a fleeting glimpse of the wealth and opportunities of their Western colleagues. The antibiotic market was worth an inconceivable $32 billion, one of the smart investors had told them. Just think how much even a fraction of that pie would be worth! At the same time, they realized that their circumstances weren't likely to change very quickly. 'We did the work. Now who is going to pocket the millions?' Meipariani asked me when I met him on my first visit to Tbilisi in 2000, some four years after the fiasco.

Despite this, the researchers retained their warmth and hospitality. During my stays in 2000 and 2002, they took care of me around the clock. For instance, when Zemphira Alavidze needed to give a doctoral exam, she just brought me along with her. And the candidate's family invited me to the exuberant party afterwards. The table featured towers of shish kebabs, salads, fruit, khachapuri (bread filled with creamy cheese), khinkali (a type of well-seasoned, oversized ravioli) and wine, of course, of which the Georgians are so proud. The festivities lasted for hours and toasts were raised at regular intervals. There was no holding back Tato Gabisonia, the man with the three jobs, while gaunt Amiran Meipariani ate silently and stood up every now and then to give a short speech. I had to stammer a few words, too. The women sang Georgian songs. The highlight of their hospitality was when a friend of the family I had never met before presented me with a rare copy of the Russian edition of d'Herelle's book.

Celebrations such as this allowed a peek into the interior of the flats, which looked amazingly good compared to the exterior of the buildings. Apart from the newly constructed mansions of the nouveau riche, the houses tended to be dilapidated ruins, with balconies hanging on for dear life to the façade on each floor. The front doors to the estates were no longer there. When you looked in from the street, you peered into dark caverns, since the lighting in the stairwells never worked and letter boxes were a rare occurrence.

Georgians improvised their oases in the crumbling state that barely took care of its citizens. The dance for power and influence was more important. Georgia was considered to be one of the more corrupt former Soviet republics. When Edward Shevardnadze, a familiar political figure in the West, was still president of Georgia, he did once manage to make his way to the Eliava Institute on Gotua Street. The reason for his visit was the very *New York Times* article that had been the inspiration for Alfred Gertler's trip to Georgia. If the state's own prophets in Tbilisi were held in such esteem by journalists in distant New York, then at least a visit was called for. Shevardnadze politely and patiently listened to the lecture given by Chanishvili, and then he disappeared in his Mercedes, a gift from the German government. The time and energy spent on this trip didn't pay off: not a penny more was earmarked for the institute by the Georgian government.

## A clever bandage brings about hope

Invaluable help in this dire situation came from Betty Kutter. With her own and donated funds, she started a foundation called PhageBiotics in order to support the Eliava Institute. For instance, the foundation paid (and still pays) for a two-year scientific training programme for Georgian students in the institute. Moreover, Kutter started scientific collaborations with the Eliava researchers and helped them to forge additional relations with other foreign scientists. The institute was also able to secure some of the money that was beginning to flow into the region from Western governments – especially the US and NATO – who were afraid that the unemployed and down-trodden researchers might migrate to suspected bioweapons labs in Iran or North Korea.

One collaboration started by some Eliava scientists and later supported by outside partners turned out to be especially fruitful. Zemphira Alavidze, who had developed the phage spray for the Georgian army, teamed up with chemist Ramaz Katsarava of the Technical University in Tbilisi. Katsarava is a high-flyer who prospered even in the difficult period after independence. He seems to magnetically attract Western funding. His success is based on his reputation as a formidable scientist and his vision that led him to look for contacts in the West when it was still risky to do so. 'Without these connections, which blessed me with house calls paid by KGB agents, I wouldn't have any research funding from the West at all today', he told me in 2002. At the time, Katsarava was a rare example of a Georgian who knew how to negotiate in both worlds. During a visit to an outpatient clinic where Katsarava and Alavidze's joint product was tested, I made a remark about the posh clinic, which struck me as outrageously luxurious. Katsarava responded laconically: 'Well yes, by the standards of today's Georgia it is in fact.'

Katsarava and Alavidze named their promising product Phagobioderm. Its basis is a clever bandage invented by Katsarava. It's made from a material that is a good substitute for skin during wound healing; enzymes are added to it so that the bandage slowly and harmlessly dissolves while it is on the wound. The bandage can be impregnated with additional materials such as analgesics and – phages. This special dressing is especially useful in treating stubborn wounds, such as those that often trouble diabetics or people with circulatory problems.

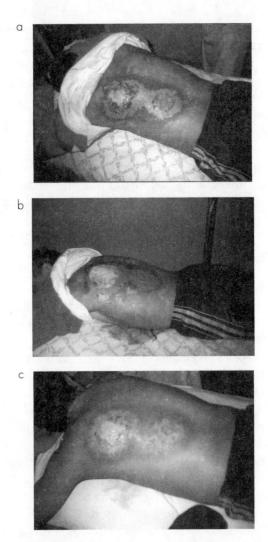

**Figure 6.2** Use of Phagobioderm in one of the victims of the December 2001 radioactive accident. The wound didn't heal until Phagobioderm was applied (visible in Figure 6.2b)

A terrible accident in a forest in northern Georgia had a small silver lining: it was a PR coup for Phagobioderm. On a frosty December night in 2001, three lumberjacks found two strange metal cylinders that were warm and melting the snow they were lying in, despite the bitter cold. The three men used the cylinders as hot water bottles. It was a horrible mistake. The containers were radioactive remnants from the arsenal of the Soviet army, which had used them as sources of energy. They burned terrible wounds in the men's skin. This incident caused secret services all over the world to jump to attention. Since 11 September 2001, these radioactive cylinders, which were found near the border to Chechnya, have been considered to be potential weapons for terrorists who could use them to make dirty bombs. The story appeared in the press worldwide, and the reports included the fact that the workers' infected wounds had been treated with Phagobioderm, because the antibiotics that were first used had failed.[7] Phagobioderm, which also provided relief for Alfred Gertler and his foot, has been approved for use in Georgia.[8]

## Worlds apart

However, after 1991, the breakdown of the institute was so severe that improvements were gradual at most. Alavidze, Meipariani and their colleagues still had only scant means to do research and brew their emergency phage medicine. They had a few glass flasks and used cans for water baths. They didn't even have gloves despite the fact that they worked with dangerous bacteria. The animal experiments that tested new batches for bacterial toxins in Soviet times were inconceivable now, because they were too expensive. Despite this, their products continued to be popular in Georgia. Phage therapy

**Figure 6.3** Phage research at the Eliava Institute in 2002.
Some lab equipment dates from Eliava's time as director,
while some is more modern

remained alive. The people on the street were familiar with bacteriophages and asked for them at the chemist or when they went to the doctor, if they could scrape together the four to five laris per box – the equivalent of 10 rides on the dented minibuses. Some Georgians didn't (and still don't) travel to their dachas without making sure they had Intesti-Bacteriophag in their first-aid kits. The mixture of over 20 phages offered protection against all local microbes that might give the stomach a hard time.

Many doctors continued to be adherents of the old therapy. One of them was Ruben Kazarijan, who I met in 2002 at the Genesis outpatient clinic, which was run by the Georgian relief organization of the same name. Compared to the run-down government hospitals, the practice, which was financed by donations, was spotless and modern. When we visited the unit, Kazarijan presented his patient Lasha, a 14-year-old orphan who had played with petrol and burnt both his legs. When relief workers found him in a government hospital, his legs were slimy, covered with scabs and pus. At the time of my visit, Lasha had been at Genesis for two-and-a-half months. Kazarijan pulled back the cover. The wounds had healed up and the scars looked amazingly good. Was it the work of phages? He answered: 'For this terrible case, I used anything that worked, including phages.'

When Kazarijan talked to me about the merits of phages, there was no stopping him, although the interpreter could barely keep up. Kazarijan used the viruses to treat tonsillitis or boils in both adults and children. Although the patients in the waiting room were getting impatient, the small and wiry Kazarijan kept going strong, using his hands as props. He had had especially good success with babies, he said, saying that in the case of chronic tonsillitis, he dropped the phage solution directly on their tonsils. The virus attack often made the accompanying inner ear infections disappear as well. He emphasized that he didn't use antibiotics and never saw side effects.

**Figure 6.4** Small-scale phage production in 2002: a colleague in Zemphira Alavidze's group pours the phage solution into glass vials. If it didn't become cloudy after three days, it was sterile and was made available for purchase

At the Georgian Burn Centre, director Iasik Beshvili offered his services to two worlds – a rich and a poor one. The government-run part of the hospital housed the burn patients from Georgia, while the renovated wing was reserved for private patients. During a visit to the centre, the first thing Beshvili did was lead me to the entrance of the room that served as the intensive care ward. Four pairs of eyes looked slowly towards the door. Neither the two children nor the two old men moved even a limb of their singed bodies, and they made no sound. The room was filled with pain. A nurse stood helplessly in front of the beds, otherwise, the room was empty. The plaster was crumbling from the walls and there were holes in the floor. Outside in the corridor, the patients' relatives were waiting. If they could afford it, Beshvili sent them to the chemist to buy phages and other drugs. 'We don't get any money for the hospital from the state', he lamented, 'and the people don't have any money for medicine either.' Beshvili swore by phages. He showed me photos of a boy who was completely covered with scars. 'Eighty per cent of his skin was burnt. No one thought he would survive. There was a great danger of wound infections and blood poisoning, but phages saved him.'

Only two doors down was the entrance to a different world, one that was new and dazzlingly white. 'Where time rules, the art of make-up or physical training fail, so rather often spiritual balance and recreating or recovering health requires surgical interference', claimed an advert in the English-language newspaper *Georgia Times* I had read a few days earlier. The paper was full of such ads for plastic surgery. They targeted the wives of foreigners working in the pipeline business or consultancy sector. In particular, those physicians who had been able to get hold of a certificate for continuing education had discovered the needs and desires of these wealthy paying customers. Beshvili and his son eagerly showed me a photo album that attested to their accomplishments: breast enlargements, smoothed out foreheads, facelifts. 'Everything is done at one-third the Western rates', said Beshvili. If needed, he used phages from the Eliava Institute to support the inexpensive healing process, he told me.

## Miracles wanted

The research ties to the West and local doctors like Beshvili and Kazarijan helped the institute to slowly recover some of its lost resources. In some labs, PCs had replaced the old Robotron typewriters from former East Germany. The most conspicuous signs of the modest successes were the heavy steel bars protecting the rooms where the busy scientists worked. Iron gates protected oases of scanty wealth in the desert of ruined institutes. In the Eliava Institute, two of these precious PCs and a lab machine had been stolen from unprotected rooms. Some especially bold thieves made off with all the telephone cables in the building. In the Tbilisi institutes, at times the only places with electricity were these rooms with bars; neighbouring rooms and the corridors remained dark. Iron bars and electric light were signs of success.

This success was at risk, however. The biggest obstacle was the country's state of emergency. Shevardnadze, the 'White Fox', as the former president of Georgia was referred to, was unable to drag his country out of the mess. His government's power barely reached beyond the borders of Tbilisi. The powerful Russians constantly threatened to intervene in Georgia, asserting that Chechen terrorists were operating there. This led to an unstable situation that damaged the country on a long-term basis. Shevardnadze himself only just survived two assassination attempts, and foreign business-people were occasionally kidnapped for weeks at a time.

Added to this was corruption, which was widespread even in the president's family. Some of Shevardnadze's ministers were allegedly involved in the kidnappings. 'A fish starts rotting at its head', Ramaz Katsarava told me at the time. 'In the past years, corruption has got worse and worse. It has entrenched itself at every level of society.'

No wonder the economy and the people were suffering privations. 'Of all the countries that once belonged to the USSR, we're at the bottom of the heap', lamented Maia Mgabolishvili, head of the Genesis relief organization. It wasn't that long ago that Russia sometimes turned off its neighbour's electricity and gas to put political pressure on the country or simply collect unpaid bills. The Eliava Institute also experienced periods without electricity because it had run out of money. Yet most researchers who had stuck it out at the institute since the problems started in 1991 continued to stay. They were by no means only veterans like Meipariani or Chanishvili. There were also young researchers who had the necessary know-know to leave for a Western lab. Their loyalty came from their love for their country, which has a long history of invasion and oppression. After all, the small country had been repeatedly invaded by Muslim powers from the southeast hungry to expand, as well as the tsarist empire in the north. Yet Georgians are proud of their country, their perseverance, their tolerance towards Muslims and Jews, their culture and their wine – said to be the oldest in the world. Every weekend, the ancient Orthodox cathedral in the former capital of Mtskheta attracts hordes of visitors who dream of the roots of their country there – and believe that miracles can happen.

In fact, the country could have used some miracles. Mzia Kutateladze from the Eliava Institute, ever the rational scientist, told me, however: 'We have to help ourselves. It's especially important that specialists stop fleeing the country.' When a foreign businessman had tried to recruit young students from the Eliava Institute, she had made their objections clear: 'Stop it! We don't want to go. We want to help Georgia.' Her reaction called Eliava to mind, who had rejected a position offered to him in France and returned to his conquered homeland.

## Tarnished treasures

Despite the experience with GRI, scientists from the Eliava Institute knew that they needed to do everything in their power to broaden their ties with their counterparts in the West. Fortunately, interest in phage therapy in the US and Europe is growing. Nearly every new company that has decided to take a gamble on this alternative has contacted the institute in Tbilisi. Western companies have their sights set on two treasures: the phage bank (probably some 3000 types of viruses are stored in the refrigerators) and the scientists' expertise.

'The expertise there is immense', says Sergey Bujanover of the Israeli company Phage Biotech. 'In the entire West, there are only 40 types of phages that are active against *Pseudomonas aeruginosa*, a bacterium that causes huge problems in burn victims. In Russia and Georgia they have hundreds of them! It's the fruit of 70 years of work and experience. It will take a long time to catch up.' Betty Kutter seconds that: 'Eliava researchers could test their phages on humans right away, which allowed them to constantly select the most therapeutically promising from the large number in nature and develop them into new drugs. Western companies don't have this option.'

Yet both treasures continue to be difficult for the Eliava researchers to commercialize. The Indian company GangaGen invited the young lab supervisor Mzia Kutateladze and a colleague to the high-tech boom town of Bangalore for the purpose of exchanging East–East know-how. 'We showed them everything: how you isolate and test phages, and so on', Kutateladze explained, sitting on the sagging sofa in her cubicle during my visit to the institute. 'Then the cooperation stopped. They wanted our advice, and they got it', she added, without a touch of anger or bitterness. The only thing she had to show for the visit was a few snapshots of the Asian boom town.

GangaGen was also interested in the hundreds of research reports that Georgian and Russian scientists had published over the years. Such publications are the hard currency of the scientific information 'stock exchange', but Eastern currency is barely worth anything. In the West, hardly anyone can read Russian or Georgian, and no one is interested in learning them. 'We're supposed to have everything translated, but who's going to pay for it?' Kutateladze asked.

The publications are not only written in languages that are incomprehensible to the crucial people – those with money or influence – but they often don't meet Western standards. This also applies to the more recent papers published long after the war: untreated control groups that serve as a comparison for the efficacy of the phages are missing in one, while in others, exact details are not provided. For this reason, at best, they provide indications of the success of phage therapy, but not proof.[9] This is why many scientists are sceptical of phage therapy. Researchers who are newcomers to the discipline, like Bujanover, agree that more careful studies need to be carried out, but do not share the general scepticism: 'The long years of experience show that it works – in exactly the right niche: for multi-resistant bacteria.'

The required studies are extremely expensive, which puts them out of reach of Georgian or Russian means. That's also something that Betty Kutter wants to help address with her PhageBiotics Foundation. It has started a new surgical infections and phage therapy training programme under Guram

Gvasalia, medical director at the Tbilisi central hospital, where phages are routinely used in surgery. One goal of the programme is to finally sort out and publish the results of some of Gvasalia's 30 years of clinical experience in treating diabetic ulcers, serious wounds and bone inflamations – the ailment that plagued Gvasalia's patient Alfred Gertler. 'The publication of Gvasalia's results could help to get the massive funding that is necessary for clinical trials of phage therapy', says Kutter. 'Such formally conducted trials are urgently needed. It is high time to start with them – but that will require outside support.'[10]

## Phages rearm bacteria

Like the Soviet-era publications, the legendary phage bank also induces ambivalence in the West. Granted, its sheer size is one of a kind, and the number of phages with which doctors have had experience is immense. But, by today's standards, the viruses haven't been investigated well enough. It's like trying to work out what's in the enigmatic black box – 3000 phages in this case. Drug approval authorities like the FDA will probably insist on decrypting the phage genomes. For Western labs that's no big deal, but for the people from the Eliava Institute it's far too expensive to do on their own.

Still, light has to be cast into the black box, because some phages can occasionally arm a bacterium with an additional poisonous gene and turn it into a lethal weapon. Scientists call these 'temperate phages'. These are the viruses that stumped Bordet, d'Herelle's adversary, when he thought that phages were enzymes produced from the bacteria. Their insidious life cycle wasn't decoded until the 1950s.

Once they have entered a bacterium, temperate phages don't always turn their victims into a phage factory right away.

Sometimes they insert their own hereditary material into that of the bacterium and hibernate in the infected bacterium, which gets the following benefit from the strange symbiosis: the virus inside the bacterium makes it resistant to phages of the same type. When the infected bacterium grows and divides, both daughter cells get a copy of this Trojan Horse. The phage spreads quickly in a population and along with it any dangerous genes it may carry.

In 1996, US researchers Matthew Waldor and John Mekalanos made this surprising discovery. Two components of the cholera bacterium help it to achieve its full virulence. One helps it to attach to the human intestinal cells, and the other is a potent poison, the cholera toxin. The toxin sneaks into the intestinal cells and manipulates the water balance there. As a result, the affected cells pump massive amounts of water into the intestine, and if the patient isn't treated immediately, he or she becomes dehydrated. Of all things, the cholera toxin is transmitted by a temperate phage that apparently prefers to do its lethal business in the intestine.[11]

The situation is similar with STEC. The acronym stands for a bacterium of the species *Escherichia coli*, which has taken on additional poisonous genes in its hereditary material. This turns the harmless intestinal inhabitant into a microbe that can bring about diarrhoea and enteritis with potentially fatal side effects. One of STEC's microbial potency enhancers is the Shiga toxin 2, which is also spread by a temperate phage.[12]

Mekalanos was once even able to trace how a temperate phage carried out its insidious genetic business.[13] A diphtheria epidemic broke out in Manchester, a rare occurrence these days. The epidemic was triggered by a child who had returned to England from Africa. Toxin-producing diphtheria bacteria had lodged in the child's throat. Once he was back in England, he infected people whose throats were colonized by harmless diphtheria bugs. Phages in the African bacteria exploited the sudden proximity and leapt across to the English bacteria, which now also hosted the dangerous toxic gene.

In addition to temperate phages, sometimes phages that don't settle in the hereditary material – called 'virulent' or 'lytic' in scientific jargon – can transmit genes. This occurs because the hasty phage production never works perfectly in the kidnapped bug. In the 'factory', phages are also produced without a tail or without any genetic material. Defects also occur when the hereditary material is packaged. Sometimes bacterial DNA is packed into the head of a phage instead of the phage's own DNA. This is then sneaked into another bacterium by this phage. This promiscuous process, which goes on all the time in the natural world, happens by chance. Usually it doesn't cause much harm. Sometimes, however, toxin genes can be transmitted to previously harmless bacteria.

If the hereditary material of a phage is decoded before the phage is used for treatment, researchers can detect any toxin genes or ones associated with the 'temperate lifestyle'. During the Soviet era, analyses that would have revealed these results were too complicated or technically impossible. 'But due to a lucky circumstance, the risk that comes from temperate phages never had an effect', says Betty Kutter. 'The Georgian researchers strictly avoided temperate phages because these viruses didn't sufficiently decimate the bacteria and as a result, the chances of healing would have been jeopardized.'

## A love finally gleans rewards

In the past few years, Eliava researchers have established a growing number of collaborations with scientists from such prestigious institutions as Rutgers University and Rice University. Some of these projects were devoted to decrypting the

DNA of several phages used for treatment. These genome sequences showed that all the tested therapeutic phages are of the virulent type.[14] An agreement that Kutateladze, Alavidze and Kutter worked out in August 2005 with scientists from the British Sanger Institute promises a big step forward along this path. The Sanger Institute is one of the leading institutions worldwide for decoding DNA; as such it was one of the important players in the sequencing of the human genome. Now the institute and the Eliava scientists plan to tackle the genomes of the major phages contained in their cocktail Pyophage, which is used to treat external infections caused by staph, streptococci, *Pseudomonas* and *Proteus*.

Despite past difficulties, the achievements of the Eliava people are now visible in their steadily improving institute. Today, many of the labs have been renovated and the benches feature new equipment like centrifuges, spectrophotometers and computers. New life is also flourishing in the production area, where different groups from the institute are building a new facility to manufacture their phage cocktails for the Georgian market. They have also signed an agreement with entrepreneurs from the US to use their products to treat patients at a Mexican hospital. In another part of the institute's main building, phage medicine production was restored a while ago. Cocktails with the familiar names of Pyophage and Intestiphage are being churned out by a company called Biochimpharm. They are sold all over Georgia and in some neighbouring republics like Armenia.

Biochimpharm is a good example for the way many of the groups from the Eliava Institute got back on their feet. The head of the small company is Alexandr Golejashvili, a former student of phage veteran Liana Gachechiladze. Together they were able to obtain financing for a basic research project and refurbishment of Golejashvili's production facilities from the

International Science and Technology Center (ISTC). This is one of the international agencies that support research in the countries of the former Soviet Union. When the first round of the project was over, the ISTC agreed to extend financing until 2006. This and the money pouring in from phage sales help Golejashvili's company to continue working on manufacturing products like phage tablets for *Salmonella* infections.

In the meantime, Liana Gachechiladze has gone on to develop a phage drug for the Western market. Quite a prospect after the disappointment of the GRI affair – even though the concoction she is working on is intended for infected canine ears and not those of humans. In collaboration with a veterinarian from the US, Gachechiladze is targeting infections with pseudomonad, the bacteria she has studied for more than four decades. As a young scientist in the 1960s, she was proud when she received an award by the Soviet ministry of health for producing the first phage cocktail to fight *Pseudomonas* infections. Today, it would be a boon if her new phage cream could sometime in the future be used on pets.

Happily, the progress made by the people from the Eliava Institute is accompanied by an improvement in the political situation in Georgia. In November 2003, Shevardnadze was ousted from power. An overwhelming majority of Georgians elected 37-year-old Mikheil Saakashvili as their new leader. His experience studying law for several years in the US convinced many voters that he would be able to change things. In fact, Saakashvili has managed to stabilize the power of central government against rebellious provinces. 'His government has also dismantled many bureaucratic hurdles', says Mzia Kutateladze. 'But there are still so many things that have to be done for a better future.'

The people at the Eliava Institute are continuing to work hard for that better future. But anyone who has seen them going to their labs even in the harshest of times knows that it is more than a desire for prosperity that keeps them loyal to phage therapy. One day during my first visit in Tbilisi, I was sitting with 76-year-old Amiran Meipariani in his bare office. A cigarette in his mouth, his clear, light blue eyes gazed through the window towards the Mtkvari River where Eliava had discovered the river's enigmatic power to dissolve bacteria over 80 years ago. 'I love phages', he professed dreamily. 'They're like beautiful women – how could I forget them?'

# 7

## resurrection

Deep inside his body, the branched tube leads from his belly to each of his thighs. To the rhythm of his heartbeat, the blood pulses through the ribbed plastic aqueduct and takes vital oxygen to his legs, feet and toes. Without the artificial artery, 66-year-old Hermann Kläfker would have already lost his legs. And all this after years of hard, physical work at the potassium mine near the German city of Hanover. Before that he had worked as a stonemason in the summer and a butcher in the winter. Ruptured disks put him on disability pension at the age of 58, and he nearly died of heart failure when he was 62.

Then there are the clogged arteries. That used to automatically mean amputation, but today artery replacement is routine. At the Medical University of Hanover (MHH), the vascular surgeons replaced a key artery, which had been clogged with six decades of sludge, with a plastic tube. They inserted a bifurcated tube, whose branches connect the arteries of the legs with the aorta. Two weeks later, Kläfker suffered such a violent asthma attack that he needed to be ventilated. In order to suppress the attacks, the doctors had to give him such high doses of cortisone that the sutures from the operation couldn't heal. Wound secretion collected in the incisions and they burst. At this point, the doctors removed the dead tissue from the wounds, irrigated them, dressed them and hoped the deep incisions would heal.

For two months, the blood has been pulsing through the vascular prosthesis, but Kläfker is still stuck in the hospital. The high-tech medicine worked. That's not what's keeping him from going home to his wife and his dog. Bacteria have chained him to his hospital bed. After his asthma attack, pseudomonades infiltrated the surgical wounds on both thighs. As long as they are there, the wounds are unlikely to heal up. And that's not the worst: 'The bacteria often don't stay on the surface', says Maximilian Pichlmaier, Kläfker's surgeon. 'They make their way down into the depths of the wound, reaching the prosthesis. Then we have to remove it, because the site of infection destroys the natural vessels where the prosthesis has been sutured into place and can pierce the intestine if it is nearby. It doesn't get any more difficult than that.'

That's because the doctors don't have a safe method for sanitizing the wounds. Simple disinfection with caustic agents disrupts the healing process, and irrigating the wound with antibiotics doesn't help either. Kläfker's bugs are resistant. In order to hinder the bacteria in their offensive towards his legs, Pichlmaier placed two tubes in the wounds which were hermetically sealed with bandages. The drainage ends in a vacuum pump. The sucking sound chugs relentlessly in the room. Liquid wanders from the leg tissue through the wound and into the tubes. The steady stream is supposed to stop the germs at the surface of the wound and perhaps even suck them away. This works in many patients, but not in others. Then the doctors have to capitulate in the face of the microbes, despite the machines, technology and drugs.

For Kläfker, the battle that is almost lost means constant bed rest. In order to keep the resistant bugs from spreading, he's in solitary confinement. No one knows how long it is going to continue. The pump, which sucks his body juices out of every bit of his body, makes him thirsty all the time. 'My family comes to visit me every other day', says Kläfker. 'It's too far for them to make the 100-km trip every day.' The TV costs €6.50 a day. 'I can't afford it. I'm going to be in here for a while.' His wife brought him a tiny black-and-white television. Since he was first admitted, not a single ray of sunlight has made it into his room, because the building next door blocks the sun's path, which is low at this time of year. It's something that Kläfker doesn't talk about unless you ask him, and when he answers, his voice is low and matter of fact. 'It's just when I see the other patients going home.' His mouth begins to quiver and tears well up. 'I don't want to stay here for many more months.' Christmas is on the way.

For surgeon Pichlmaier of the MHH's Department of Thorax, Heart and Vascular Surgery, Kläfker's fate is not unusual: 'This is where all the difficult cases show up that the smaller hospitals can't handle or don't want.' Pichlmaier and his colleagues at the MHH are confronted with the might of microbes every day. They defy drugs and destroy the precision work with vascular prostheses or artificial heart valves. 'In view of the resistance problems, we need an alternative to antibiotics', says Dieter Bitter-Suermann, the retired director of the Department of Medical Microbiology and Hospital Epidemiology at MHH.

His colleague Pichlmaier fights the bugs' insatiable hunger with his non-stop activity. There's no time for anything super-fluous in his schedule. He is constantly performing operations. An inline scooter helps speed his travel along the endless corri-dors of the MHH. His office, a cubbyhole, is completely stuffed. Pichlmaier seems to find the rapid pace of his explanation, which he has spiced up with a series of queasiness-inducing slides, too slow to contain his thoughts. 'This is a piece of vascular prosthesis that has been removed', he says quickly, pressing a clear plastic container, the kind you get in a salad bar, into my hand. Reddish liquid is leaking out of the crack between the container and the lid. 'Is that blood?' I ask meekly. 'Yes, but this one here is way worse.' Now he produces a new container. This time the tube is floating in brownish iodine tincture. 'See the slimy coating on the prosthesis? That's the resistant bacteria.'

Difficult cases abound that no one outside the hospital walls hears anything about. They are fates that most people assume belong to distant times, thanks to antibiotics. It takes Pichlmaier just a few minutes to break this illusion. He tells the story of a drug addict who ended up on his ward as the result of an injection abscess. The pus-producing site had riddled an artery in his right thigh with holes. Pichlmaier repaired the vessel with a piece of one of the patient's veins. 'But the piece we inserted simply melted in the infection.' The surgeon had to replace the destroyed vein transplant with one from a donor. In the meantime, however, the opening on the thigh was infected by multi-resistant pseudomonades. His leg swelled up, making walking impossible. A slimy green, putrid film spread out on top of the gaping wound. 'If you were to rub it off with a brush, it would reappear the next day.' At this rate, the wound, which revealed the muscles and other anatomical details and was now several weeks old, would never close up. What could be done? Another patient, who had suffered burns on his arm, shoulders and chest six years earlier, didn't even bother to answer such a question. Red wound islands floated in a sea of scarred skin. Resistant staph, pseudomonades and *Proteus* bacteria colonized his burn wounds. The pus-covered film gave off a horrible smell. The 27-year-old man had already avoided people for a long time. Like an outcast, he remained withdrawn in his room, alone and cut off from society.

A businessman who had burned his buttocks on an oven was tormented for 20 years. The plate-sized wound wouldn't heal because multi-resistant staph (MRSA) had settled there. Where normally the skin protects the thighs and buttocks, raw flesh bulged out. Sitting in a normal position was out of the question for this man. A makeshift solution with a special chair and a sitting position that called for him to lean far forward over his desk provided a bit of relief. This went on for two decades. Medicine had nothing to offer him.

For Maximilian Pichlmaier, that's not enough. When conventional methods fail umpteen times, alternatives are called for. The surgeon sees bacteriophages as one of these alternatives. The next thing he wants to do is set the microbes loose on Hermann Kläfker's wounds, so that he can finally get out of the hospital and go back to walking his dog. 'The first thing I'm going to do when I get home is sit in the bathtub for two hours', Kläfker says. 'I'd give anything to be able to do that.' But first he signed a consent form allowing Pichlmaier to use phages. In many countries, such 'compassionate treatment' is permitted in cases where nothing works and the patient is willing. 'These are the cases where the damage has already been done', says Pichlmaier.

## Phage lavage in the opened abdominal cavity

Where did Pichlmaier even hear about phage therapy? It's been 50 years since Polyfagin and Asid were used in Germany. The culprit is Nodar Danelia, a doctor from Georgia. He's responsible for establishing the crossroads in Hannover where East and West, high-tech and old medicine meet. Danelia came to Germany 12 years ago and worked in the Department of Trauma Surgery at the MHH. In the Soviet Union, he had used phages to treat countless patients. Once in Germany, however, it didn't occur to him to use the healing viruses. 'Then there were more and more cases of MRSA', Danelia recalls. 'And I thought, you know what, we have to do something.' For a year, he urged his colleagues at the hospital to try out phages. His efforts were in vain until an emergency situation arose in the department. 'All of a sudden six patients with MRSA appeared on the ward. No one knew what to do', Danelia relates. 'I ordered some phages from Georgia, and we used them for the first time.' After this act of desperation, in 1999, Danelia and his colleague Burkhard Wippermann treated other patients in emergency situations, including the burn victim with the stinking wounds and the businessman whose wound had made sitting in a regular position impossible for 20 years.

Of the nine patients treated with phages, seven of them responded to the therapy. After six years of living in exile, the smell had disappeared from the burn victim's wounds in three days because the ranks of bacteria had been seriously decimated. After another 48 hours, the doctors could no longer find any microbes in the wounds, and one month later they were almost completely closed up. The businessman's wound also became sterile and healed up after phage therapy. In two patients the phages failed, however.[1]

For Nodar Daniela, the success comes as no surprise. This cannot be said for his German colleagues. They still consider the effect of the resurrected method to be unproven. This distrust was underscored by the fact that the clear solution they used to irrigate the patients' wounds comes directly from Tbilisi. For all of them, courage was required to treat German patients at a German university hospital with the phage broth that Zemphira Alavidze manufactured at the Eliava Institute. It was produced to the best of her abilities but under primitive conditions.

The Georgian aid shipment violated nearly all the conditions that a Western drug is required to meet: the concentration of the effective ingredient – the viruses – varied from batch to batch. The same goes for the shelf life. In addition, the solution contained large amounts of bacterial debris that can cause the immune system to run amok. For this reason, the ethics commission at the MHH, whose permission was required, only allowed patients to be treated who were suffering from super-ficial wound infections with MRSA microbes.[2] Despite this restriction, some German virologists criticized the ethics commission for allowing the experiment to be performed at all. In their eyes, the risk of using the Georgian preparations was too great.

But the German doctors who agreed to get involved in the treatment with Danelia see the outcome as positive, even though their assessment is cautious. 'We managed to get seven out of the nine wounds sterile. That's a success in itself. It's worth continuing the research', Wippermann says. 'Phage therapy is an exciting approach', says Bitter-Suermann, who supported the trial in his capacity as a microbiologist. 'Some doctors are very interested in this type of therapy, not just at our university.'

With this positive feedback, Danelia can hardly wait to help phages make their breakthrough in Germany. He doesn't just mean using phage therapy for emergencies or with special permission. Danelia wants to achieve regular approval for the use of phages in Germany. It would be the first Western country where phages would get a second chance after their chaotic beginning.

Revival could be difficult, though. The German regulatory authorities and Danelia's German colleagues now want to see clear proof that phages work. The few emergency trials aren't enough to provide definite conclusions. For this to happen, larger and more controlled studies are required that strictly adhere to scientific rules and regulations. 'Danelia needs to systematically prove effectiveness for a certain indication in a large number of patients', says Bitter-Suermann.

As a scientist, Danelia doesn't have a problem with this demand. As a doctor with 25 years' experience with phages in the Soviet Union, however, it's hard to live with. After all, he has been an eyewitness to the magical healing powers of phages countless times. Whether it's a matter of pus-filled linings of the lungs, pneumonia or stubborn ear infections, he has seen it all heal. He told me about a man who was brought to the hospital after being knifed in a brawl. His stomach was full of stab wounds, calling for emergency surgery. 'I removed his stomach and intestines and irrigated them with phage solution to prevent an infection', says Danelia, his hands in the air. In one hand he holds an imaginary colon and with the other he washes it off with phage solution. It goes without saying that the injured man did not have an infection after the phage lavage. 'So what do you think?' asks Danelia. You can read the second, unspoken question in his face: 'Why is everyone so sceptical?' A minute later, the scientist has retained his composure and points out the problems. The biggest problem is lack of money. The expenses for clinical studies can amount to umpteen million euros.

In late 2001, Danelia resigned from his job as a trauma surgeon at the MHH. Since then he has been tinkering with his dream. And he's set up a company. The Innovation Centre in northern Germany has provided start-up assistance. Danelia is establishing a phage bank with Georgian viruses and those he has isolated himself. He has commissioned a drug lab in Hannover to manufacture purified phage mixtures that for the most part are free of dangerous bacterial debris. The DNA of the individual viruses is gradually being decoded, so that the viruses carrying potential toxic genes can be discarded.

In future, Nodar Danelia would like to follow in the footsteps of Inga Georgadze, who operates the diagnostic clinic Diagnos

90 in Tbilisi, by setting up a similar operation in Germany. The doctors who work with Danelia send their patients' swabs or blood specimens to a specialized microbiological lab. There the bacteria in the samples are identified and the matching phages selected that the doctor can use in his or her practice. The physicians play a key role in this scheme. 'You have to master the method or you'll see a high failure rate', Danelia says. For instance, many infections are not caused by a single bug but several. The doctor, in cooperation with the lab, has to be on top of this in order to select all the required phages. 'Then you have to make sure to create good starting conditions for the phages. Before treatment begins, the dead tissue needs to be removed from the wounds as completely as possible, and the wounds have to be sanitized so there are fewer bacteria and obstacles for the phages.'

For the time being, Danelia, along with some interested doctors, continues to treat hopeless cases. 'We work on those cases where doctors are stuck between a rock and a hard place, with no alternatives', Pichlmaier says. He was the first doctor to express interest in phage therapy after Danelia's small-scale study in the Department of Trauma Surgery. Gradually other doctors approached Danelia with their difficult cases. These include patients in a persistent vegetative state whose throats and lungs are infested with germs that can't be dispelled by antibiotics, or diabetic patients whose feet have poor circulation and open wounds and are on the verge of needing to be amputated. He can't complain about the lack of work.

In a break during an operation, Pichlmaier reels off some of his cases: the above-mentioned drug addict with the injection abscess whose veins the bacteria ate away, for example. After five days of phage therapy, the wounds were sterile and the doctors could finally make a skin graft. In the end the wound closed over. Then the old woman whose chest incision had been sutured shut after a heart operation, only to have microbes invade the wound. Thanks to phages, the incision became bacteria-free. Contrary to the instructions in the text-books, doctors did not have to change the wires holding the sternum together. And what about Hermann Kläfker? He hasn't been as fortunate. In lab tests, all the phages from Danelia's collection have failed. 'You're just going to have to call Georgia again', says Pichlmaier quickly to Danelia. Then off he goes to the operating room on his scooter. A patient is waiting there for him to excise parts of his foot. The microbes were the winners this time.

## 'We would rewrite the surgical textbooks'

Pichlmaier and Danelia are scraping money together bit by bit in order to carry out a scientific study. They will be testing a therapy that could help people with cystic fibrosis (CF). Some 8000 people in the UK suffer from this hereditary disease, and in the US the number is 30,000. Their average life expectancy is not much more than 30 years. A genetic defect makes many of the body's glands produce thickened secretions. This causes thick, sticky mucous to collect in the lung, clog the bronchia and make breathing difficult. Staph germs colonize the area. Later, they are joined by pseudomonades that soon become resistant, as a result of the permanent use of antibi-otics, and hide under a tough protective layer. 'These people walk around with pus in their lungs', says Pichlmaier. The stress can lead to parts of a lung collapsing or even an entire lung bursting.

In extreme cases, doctors exchange the damaged lungs for a donated organ. Transplantation often helps, because the new lung does not carry the genetic defect and, as a result, does not produce the thick mucous. For many patients, however, the new organ doesn't make any difference. 'The very first breath the freshly operated patient takes transports a huge amount of bacteria from the windpipe into the new lung', says Pichlmaier. They are given antibiotics, but the bugs are usually resistant. On top of that, the immune system must be suppressed because of the rejection response, even more than in the case of heart or kidney recipients. 'That means a free-for-all for the bacteria.' Many a donor lung never recovers from the shock and the patient dies. Phages could prevent the microbe attacks from the very start. Pichlmaier would like to use matching viruses to free the respiratory tract before the operation so that the new lung would have a stress-free start to its new life. 'That would be fantastic', says the surgeon.

At the time this book goes to press, the study should be underway. 'We already have the permission of the responsible ethics commission', says Pichlmaier. The focus of this initial phase is not healing patients but rather testing the principle of the therapy: can phages cleanse a CF patient's colonized respiratory tract of bacteria and protect a newly transplanted lung from involvement? In the experiment, Pichlmaier will treat infected lungs that have been surgically removed because they have been so heavily damaged. The doctor places the sick lungs, which would normally go in the incinerator, in a tank in order to keep them alive for another three days. He pumps replacement blood through their vessels, warms the lungs to 37 °C and ventilates them through a tube placed in the bronchus. Each lung is placed in its own glass container. This enables Pichlmaier to use the ventilation tube to treat one lung with nebulized phages, while the other lung serves as an untreated control. 'If the phages do the job, it won't be a big step to use them in trials with patients, since we're carrying out this preliminary study on human lungs', he explains.

If enough funding were available, Pichlmaier could try to deal with other problems, for example the infection of vascular prostheses, which is also threatening Hermann Kläfker. If, in cases like his, the doctors can't keep the microbes away from the transplanted bifurcated tube that transports the blood from the aorta to the legs, it becomes life-threatening when surgeons do not remove the prosthesis and the infected tissue. 'At the same time, we have to place a provisional prosthesis from one arm to the top of each leg in order to supply the legs with blood', explains Pichlmaier. 'It's a huge procedure that some patients don't survive. If we could sterilize infected prostheses with phages or could prevent infections by using phage-impregnated prostheses, it would be an enormous help. We would rewrite the surgical textbooks.'

## Polish phage therapists also want to get on the bandwagon

The places where phages could be really beneficial are easy to locate on the map of suffering. That's why Nodar Danelia and Maximilian Pichlmaier are not the only ones who want to bring phage therapy to life a second time. In Germany's neighbour Poland, competitors are trying to be the first ones to grab a share of the $38 billion antibiotic market. The group at the Institute for Immunology and Experimental Therapy (IIET) of the Polish Academy of Sciences in Wroclaw would like to turn its long years of experience in the fight against multi-resistant bacteria into cash. When the resistance situation in Poland became increasingly threatening in the late 1970s, a group, led by Beata Weber-Dabrowska and the late Stefan Ślopek, set up a phage fire brigade independent of their colleagues in the Soviet Union. In the past 20 years, Weber-Dabrowska's team has treated over 1300 patients. Their therapy has targeted a wide range of problems, including resistant *Salmonella*, staph,

*Klebsiella* and pseudomonades, purulent meningitis, blood poisoning and infected wounds. It seems as though there is no bacterial disease that the Wroclaw-based group has not accepted for treatment.

While their focus was on emergency treatment, Ślopek and Weber-Dabrowska's team tried to prove the effectiveness of their undertaking by carrying out statistical analyses. If you put all their areas of application together, their rate of success is 85.9 per cent, the scientists reported in a publication in 2000. If the applications are listed separately, the success rate starts at 61 per cent for diabetic feet and reaches 100 per cent for boils.[3] The case descriptions and analyses are not considered strictly scientific evidence because there were no untreated control groups involved. However, many Western scientists regard them as the best work that phage therapists from the former Eastern bloc have published so far.

According to Andrzej Górski, the new head of the institute, the Polish group is now carrying out controlled studies. 'But I don't have any results ready for publication yet.' Originally, Górski planned a joint venture with a company called BioTix in order to market the IIET's phage therapy know-how. Within five years, BioTix wanted a 10 per cent share of the Polish market for antibacterial drugs. So far that hasn't happened. However, the IIET was able to post a success recently. In June 2005, it obtained approval from the local ethics committee to start carrying out 'experimental phage therapy' in the case of resistant infections again. 'This approval is based on EU law', Górski says. He adds: 'We now have a small outpatient clinic with five rooms.' Is the institute going to advertise its services outside Poland? 'I don't know yet', he says. But he stresses that the demand for IIET's phages is huge and so far they have been sold at cost. 'If we continue at this rate, we'll be broke in no time.'

## 'Rambo cures rats, mice, rabbits and pigs'

The fact that several companies are in the race is to patients' benefit. The hurdles are high for new drugs. On the journey from lab to clinical tests, with the ultimate goal of the pharmacy, the majority of all potential drugs fall by the wayside. The more companies and universities that work on a new method, the higher the chance that at least a few will be successful.

In the US in particular, a whole series of companies would like to be the ones to launch the first phage drug on the market. And there are good reasons for this. Although the population of Europe and the US is nearly the same, Americans spend twice as much on drugs as do Europeans, and the tendency is rising. In addition, the prescription-happy approach of US doctors has made the resistance situation there as threatening as it is in the UK. Multi-resistant bacteria have become entrenched in many hospitals in both countries. In response to this crisis, nearly a dozen start-ups in the US, Canada, the UK and India are doing research. They are joined by projects being carried out by research teams at universities. Yet after the disastrous start of Caisey Harlingten and Richard Honour in Tbilisi, most of the researchers are keeping their distance from Stalin's legacy.

The idea is to reinvent phage therapy from scratch, free of the baggage of the wild pre-war years in the Eastern bloc. Advocates like Richard Honour think this is the only way phages will have a chance to return to US doctors. 'In the eyes of the scientific establishment, phage therapy still has a bad name', says Janakiram Ramachandran, founder of the US-Indian phage company GangaGen. The mistakes of the tempestuous pioneers and the flawed work of Soviet researchers remain a problem.

The untested explanations of some critics also contribute to the bad image. In 1963, influential phage researcher Gunther Stent, who belonged to the Phage Group, explained why phage therapy didn't work in his book *Molecular Biology of Bacterial Viruses*. According to him, the body's immune system finishes off the phages, stomach acid destroys the phages that have been ingested and the bacteria become resistant to the phages. Back then there were enough antibiotics, which meant that it wasn't necessary to follow up these suspicions in experiments. Yet the claim had been made and for many infectious disease specialists it remains valid today. 'That's why we have to proceed carefully now and convincingly prove the safety and effectiveness of phages once and for all. Otherwise we'll never win over the establishment, and the method will perish. And that would be a real shame, because it works', Ramachandran says.

GangaGen head Ramachandran is part of the research establishment himself, but he's too modest to mention it. Before he retired in June 2000, he worked as a director of R&D at the research centre of the pharmaceutical giant AstraZeneca in Bangalore, India. After he retired, he was convinced of the merits of phage therapy and founded GangaGen. He was able to recruit a number of distinguished researchers for GangaGen's scientific advisory board.

Richard Honour has experienced the significance of the establishment's opinion in times of volatile stock market prices and stranded start-up companies. After his return from Georgia, in posh labs the researcher started developing a variant of phage therapy that could be marketed in the US. In summer 2002, he happily led me through the small company headquarters in Bothell, a suburb of Seattle, Washington. In a mixture of an infomercial and mysteriousness, which is characteristic of the biotech industry, the jovial company head hawked his remedy:

'Rambo, our number one phage, has a very specific appetite for multi-resistant staph. We've tried it on the world's worst villains, and it killed 98 per cent of them.' Yanking open a refrigerator, he said: 'See, 3600 bacteria are stored in here. We've collected them from patients all over the world who succumbed to their infected wounds.'

'Where did you find Rambo? In sewage?'

'I can't tell you. But it doesn't come from sewage. That sounds too disgusting for the patients and the FDA. Rambo is from a very unusual source.' None of the experts I consulted after this interview could imagine what Honour could have been referring to.

'How far along are you with your tests?'

'Rambo cures rats, mice, rabbits and pigs that have blood poisoning when we inject it intravenously – without any side effects. We've developed secret purification steps that reliably remove the dirt. See!' He took a bottle filled with a clear, viscous fluid out of the refrigerator. '$10^{11}$ high-purity phages per millilitre, absolutely safe. I've already drunk some of it myself.' Apparently self-experiments are still popular today.

'Where are things going from here, Mr Honour?'

'We'd like to begin testing the phages in humans who have staph-induced eye infections. It's a small market, but the victims can go blind. Antibiotics take days to reach the eye at the needed concentration. When you put phage drops in the eye, however, they start working immediately. It will take me only five months to get the documents together and submit them to the FDA. Thirty days later I can then start doing clinical trials – if I can raise money again.'

Honour was amazingly upfront when it came to this critical point:

We've been struggling for five years. During this period we had enough cash for about 30 months. At the moment the entire biotech sector is down the drain. We held our own pretty well, but right now we can't pay any salaries.

Six months later, in late 2002, he had to move his company into smaller labs. A few months later the start-up folded, and Honour founded his new phage company Viridax. The man who 'shattered the dream' of the phage researchers in Tbilisi seven years ago is himself fighting a lack of money.

## A milestone: a clinical phage study in the US

The curse of the money drought is a real burden for the sector. It makes desperate company heads brag about their results, like Honour did, while others have even been shown to have lied outright.[4] The financial crisis has also caused many promising projects to fail. This is what happened to Exponential Biotherapies of all companies, which was the first and thus far only US company that had received official approval for a phage study in humans. Three years ago founder Richard Carlton was completely optimistic. Despite secrecy that made Honour's tight-lippedness seem small by comparison, he even allowed a personal interview to be held – on his own terms. This did not include a visit to the company's lab in Rockville, Maryland, however. No journalist and no interested researcher had ever been allowed to visit the lab. Instead, Carlton's invitation was to Port Washington on Long Island, 400 km away, where he ran his company. The meeting was at a hamburger restaurant at the train station, since his office was also off limits. He said he could afford to talk for only two hours, not a second longer. 'We're concentrating on a drug for enterococci', Carlton said. They are the bugs that normally live peacefully in everyone's intestine, but as multi-resistant variants, they plague patients with weakened immunity. They colonize the bladder, the blood and the heart and are almost impossible to stamp out. Enterococci lead the pack of resistance artists. Strains that are resistant to vancomycin, the antibiotic of last resort, have been around for a long time. This earned them the abbreviation VRE: vancomycin-resistant entercocci. In US hospitals, the situation is particularly critical. Between his cheeseburger and cole slaw, Carlton spread out the visuals, with the red curves and blue bars demonstrating what Exponential's phages could do.

They were capable of doing a great deal. They saved mice at a healing rate of 100 per cent, even when researchers injected a load of VRE in their abdominal cavities far beyond the amount a human patient normally needs to deal with. 'There's nothing conspicuous about the infected animals that we treat using phages. Their coat is a bit ruffled, and they're slightly lethargic. If you compare this to humans, it's like having a mild cold', Carlton said. In contrast, untreated mice lay doubled up in a corner of the cage after only 8 hours. A yellow liquid poured out of their closed eyes and 40 hours later they were all dead. These results were published at the time.[5] Carlton ordered a quick coffee and allowed himself the luxury of surrendering some unpublished results. In a small experiment with two rabbits, Exponential's phage even outstripped the new antibiotic synercid. The animal that was treated with phages lived, while the one that was given synercid died.

Another experiment was supposed to replicate the day-to-day emergency situation in an intensive care unit. An older patient has just received a new kidney. In order to prevent his immune system from rejecting it, drugs are used to suppress it. Despite all precautionary measures, the patient is infected with multi-resistant enterococci. Vancomycin is ineffective. The bacteria in the blood run rampant and the patient dies. The researchers at Exponential used animals to simulate this scenario. They suppressed the immune system in 30 mice with a drug and then infected them with VRE microbes. They left 10 mice untreated, and only 1 of them survived. Ten mice were given the new antibiotic linezolid; 4 survived. The remaining 10 mice were given phages that saved 6 of them. The FDA had requested this experiment because it reflects one of those crisis situations for which new drugs are so urgently needed. 'We began discussions with the FDA in 1999', Carlton said. 'They asked a lot of questions, but they were pretty open.'

They were so open that Exponential Biotherapies was allowed to proceed with clinical studies. It was a milestone on the way to the first phage drug to gain approval in the West. In a Phase I study, it was first tested whether Exponential's phage was harmful for people. Doctors injected the high-purity phage directly into the bloodstream of 30 healthy subjects, twice a day for a period of nine days. Apart from a temporary rash in one subject, they did not observe any side effects. However, the planned Phase II study to treat patients with chronic urinary tract infections never got off the ground. Carlton was unable to find a financial sponsor for this expensive step, despite the fact that leading university hospitals in the US were interested in the experiment.

## Investors want blockbusters

The high level of caution on the part of investors is fed by the FDA's broad silence on phage therapy. It has not yet published guidelines for the approval of phage drugs, but most experts expect that it will impose strict requirements for the use of viruses, which would cause the costs for developing the drugs to skyrocket. Phage researcher Tony Ilenchuk suspects that it won't be easy to sneak the phages through the maze of regulations. 'If only one of the companies fails, it's going to get tough. That's what happened when artificial blood was introduced back in the 1990s. The FDA urged the companies to go slowly back then. But one company launched a study – and some subjects died. After this incident, the FDA spent three years developing new guidelines. We have yet to see a product.'

The statements that FDA representatives have made in discussions with phage researchers provide some insight into the regulatory authority's misgivings. In addition to the reservations against phage cocktails that Richard Carlton reports, there are other issues that the FDA would like to see resolved. For example, FDA scientists are afraid that phages change in the course of production. When they multiply within the bacteria, errors continually occur while their DNA is amplified. Over time, these mutations in the phages' genes accumulate, which can have unexpected effects. 'We can't ignore these reservations', says Pablo Bifani. He spoke to the FDA several times when he was still working for the British phage company PhageGen. After the company abruptly switched to the gold business, Bifani moved to the Pasteur Institute in Brussels.

Many investors are also afraid that phages, which are naturally occurring, cannot be protected well enough via patents. Without patent protection, the high costs of development can't be recouped. There is one hurdle after another, and they are joined by the one that meant the end of Richard Carlton's phage experiments. 'Phages aren't blockbuster drugs', he says. The term 'blockbuster' refers to drugs that bring in at least a billion dollars per year. For phages, sales of this magnitude cannot be achieved, because each bacterial species requires at least one matching phage drug, and this strictly limits the market for the individual drugs. 'For this reason, we've changed the priorities', says Carlton. His Exponential Biotherapies is no longer targeting human patients, but has switched to animals. The same strategy is being followed by a number of other phage therapy companies that want to avoid the obstacles in the human medicine sector. Sergey Bujanover's PhageBiotech also halted its projects involving humans and is now aiming to use phages to improve the often ailing health of breeding shrimp. 'It's taking more time than anticipated to wean the world off the standardized wide-spectrum chemical solutions [of antibiotics] in favour of the relatively service-intensive phage technology', says Asher Wilf, head of PhageBiotech. 'This may be due to underestimating the scope of the antibiotic-resistance crisis. The fact remains that there are still very few of us and we are all struggling.'

## 10,000 cows die for no good reason

The new strategy may work because the demand for alternatives to unpopular antibiotics in animals is huge, while the supply is meagre. The jam-packed cattle pens that are characteristic of modern livestock breeding are a paradise for microbes. Whether it's *Salmonella* or *Campylobacter* in chicken intestines, or the out-of-control *Escherichia coli* variants O157:H7 in the digestive systems of cows or staph on their udders, the bacteria are just waiting for an opportunity. The danger they harbour is twofold: some bacteria attack the animals themselves, like certain *E. coli* variants that badger young calves with diarrhoea that is often lethal. Other bacteria romp around in the breeding animals without harming them but can be dangerous for people. Staking their claim to fame are the enterohaemorrhagic *E. coli* (EHEC) strains, which also include the O157:H7 bug, responsible for the 'hamburger disease'.

In the US, Canada and Japan in particular, the hamburger bug has unleashed mass outbreaks on a regular basis. In Japan,

11,000 people were stricken at the same time after they ate turnips from a field that farmers had fertilized with contaminated cattle dung. It's not difficult for that to happen. A study showed that in the US every summer, up to half the cattle harbour *E. coli* bacteria. The US agricultural authorities are now so cautious that, in July 2002, they had over 8.5 million tons of ground beef from a slaughterhouse destroyed because 19 people became ill after eating hamburgers made of beef from this source. More than 10,000 cows died for no good reason.[6]

EHEC is widespread in Europe as well, however. In 2004, over 700 people in England and Wales contracted it.[7] They were often young children or older people who had eaten undercooked beef or drunk unpasteurized milk. Ingesting 100 hyped up *E. coli* germs is enough for them to multiply in the intestine and cause stomach cramps. The diarrhoea, which starts off watery, often becomes so bloody that some patients talk about 'only blood and no stools' when they are asked to describe what they are eliminating.

The researchers do not know exactly why EHEC does this kind of destruction in the intestine, while normal *E. coli* live there peacefully and harmlessly. So far they have identified some additional proteins in EHEC that have a toxic effect on human cells. Furthermore, the bacteria have the ability to attach to the intestine. It takes 8–10 days until the body has managed to deal with the EHEC microbes and the sickness subsides – if there is no occurrence of an enigmatic complication, so-called 'haemolytic uraemic' syndrome. In this syndrome, the red blood cells and platelets are damaged, ultimately causing the kidneys to fail because of the flood of debris. Then the patients, most of whom are very young, are in a critical condition. Some of them die.

The list of scourges that strike humans is complemented by *Salmonella*, *Campylobacter* and *Listeria*. *Campylobacter* are usually transmitted by chicken, and *Salmonella* are also passed via eggs. Both of them primarily pester the alimentary canal, although severe symptoms such as muscle paralysis or shock also occur. *Listeria* usually lurk in dairy products or raw vegetables and are especially insidious because they are able to multiply in the refrigerator. *Listeria* generally cause meningitis, which develops into brain abscesses and can be lethal, especially for infants and elderly people. In England and Wales, *Salmonella* struck 13,000 times, *Campylobacter* 42,000 times and *Listeria* 110 times in 2004.[8]

Checks carried out by health authorities attest to the widespread presence of germs in our daily food. For instance, tests in the UK revealed that half of all poultry products were contaminated with *Campylobacter*.[9] That's alarming, because in the case of *Campylobacter* it only takes small amounts of bacteria for it to be contracted. Apparently, it doesn't have to be ingested in food, but infection probably also occurs as a result of contact with raw meat on a person's hands and mouth. Does this mean that in the future cooking should only be done wearing gloves? Some restaurants in the UK only serve rare hamburgers if guests are willing to sign a liability waiver if they get food poisoning.

No wonder health authorities and farmers are doing everything they can to keep bacteria out of their farmyards. In the EU alone, 3400 tons of antibiotics made their way to feed troughs and veterinarians' syringes in 1997. Yet using the drugs is becoming an ever bigger burden for the managers of pigsties and battery egg farms. Frequent antibiotic scandals leave consumers with a bad taste in their mouths. In a study published in August 2005, for example, researchers at the Health Protection Agency found *E. coli* germs that were resistant to three or more antibiotics in over half the British chickens they examined.[10]

Health authorities are faced with a dilemma. On the one hand, they have to see to it that meat is free of dangerous bacteria. On the other hand, they have to restrict the use of antibiotics in farm animals because consumers will start to rebel and the resistance crises will become more acute. Phages could offer a solution. Experts at the departments of agriculture in the US and Canada rate the potential of phages so highly that they have started phage therapy projects at their institutes.[11] Like PhageBiotech and Exponential Biotherapies, other companies are also pushing phage therapy for animals. The environmentally friendly alternative is supposed to attract consumers, and, if possible, they will familiarize future human patients with the idea that phages could be a remedy for them as well. At this point, it's still unclear how the masses will respond if doctors start prescribing viruses to cure infections. Once the phages have proven themselves in feed troughs, it may be easier to get a foot in the door of hospitals.

## Using deceit and wiliness to trick the resistant microbes

The pioneer for veterinarian phage therapy has been dead for over a decade. He was Williams Smith, a Welshman who used this method in the 1980s at the Institute for Animal Disease Research in Houghton, Cambridgeshire. At that time, no Western scientist who still had an eye on his or her career would have touched the topic of phage therapy with a barge pole. 'Back then everyone in the West was convinced that phage therapy was dead', says Paul Barrow, one of Smith's students. 'But Williams Smith was an extraordinary man. Of all the people I know, he came the closest to being a genius. Every ten years he came up with a fantastic idea.'[12]

In early 1980, this idea was phage therapy. As a bacteriologist at the Institute for Animal Disease Research, Smith was all too familiar with the dilemma of using antibiotics in animal husbandry and even then was trying to come up with alternatives. He had worked with phage typing earlier. The method, described in Chapter 4, uses phages to identify the various bacterial strains. This meant that phages were no strangers for Smith. In addition, he was also one of the scientists who was aware of the mystery of how pathogenic *E. coli* bacteria like EHEC can wreak havoc in the intestine. This knowledge helped him to design experiments that continue to be seminal today because they show how well phage therapy can work.

Smith's point of departure was an *E. coli* strain that triggered dangerous meningitis in infants. Before the trial, he had discovered that a molecule labelled K1, which sits on the surface of this aggressive *E. coli* variant, was responsible, at least to a degree, for the fierce virulence. Smith wondered whether the *E. coli* strain could be specifically kept in check if he found phages that docked onto this potent K1 molecule in order to penetrate the bacteria.

He isolated anti-K1 phages from sewage, and they indeed acted more aggressively towards the *E. coli* K1 microbes than the phage types that docked onto bacteria on other surface molecules. Instead of the usual number of one million to one billion phages, it took fewer than 10 viruses to destroy a culture of K1 bacilli. Smith started his animal trials using this super-phage. He injected 100 times the lethal dose of K1 bacteria into a calf muscle or brain of mice. Eight hours later, he injected the sick animals either with anti-K1 phages or various antibiotics. The results were spectacular. A single dose of viruses was more effective than most of the antibiotics that had been administered eight times. Only streptomycin achieved the same effect: of 30 mice treated with phages, 2 died. In the case of streptomycin, 3 died, and with the other antibiotics, the deaths ranged between 26–30 – meaning all of the mice. Around 3000 intravenously injected phages were enough to save the mice from a dose of 30 million microbes that were given eight hours to multiply in the animal. Even when he did not inject the phages into the bloodstream, but in the right hind leg, which is some distance from the left calf muscle where the bacteria were administered, 30,000 phages sufficed. Apparently the K1 phages also multiplied under these circumstances and advanced from the leg muscle to remote parts of the body via the blood. Smith found them in the spleen, the brain and the liver.

Smith didn't forget to carry out the control experiment that his early predecessors had often omitted. When he injected several mice with extracts consisting of dissolved bacteria without phages, there was no effect. 'Willie Smith had an eye for thousands of details', Barrow recalls. He also tested whether there were phage-resistant bacteria in the mice, which Stent had predicted in his textbook published in 1963. In fact, Smith did find a few resistant bacteria at the site of injection; however, they didn't appear to hinder the therapy, since he also found them in animals that recuperated afterwards.

Smith explained this with the notion that these resistant subvariants did not have the dangerous K1 molecule and thus were not attacked by phages. Instead, they behaved less virulently – like a defanged poisonous snake – and were easy prey for the immune system of the mice. He had speculated that this type of effect could occur as he was selecting the anti-K1 phages.[13] Today, researchers know that the K1 molecule conceals the underlying layer of the bacterial wall and hides it from the attack of the immune system. For unknown reasons, K1 itself is only haltingly attacked by the body's immune system.

Using this deception, Smith showed how Stent's objections could be refuted and the selection of resistant bacteria could be avoided. Phage therapists only needed to select viruses that attack the bacteria on a virulence molecule. Walter Ward's group's success was the result of a great deal of luck. As described in Chapter 4, they successfully treated typhoid fever with phages in the 1940s. Their viruses attached to the outer membrane of *Salmonella typhi* on the Vi molecule, which increases the virulence of the typhoid agent. Smith's craftiness creates an advantage for phages over antibiotics, because antibiotic-resistant microbes are normally not less virulent than their attack-prone comrades.[14]

Although he had done his first experiments with *E. coli* variants that infect humans, Smith's next goal was more in line with his job as an employee of an agricultural research institute. He attempted to save newborn calves from the fatal diarrhoea that was triggered by another *E. coli* strain. However, this time his search for phages that would attack the bacteria in question at a sensitive site proved fruitless. This led him to try out another trick. He chose a combination of two phages, and only the first one attacked the original bacteria. The second one – the 'sweeper' – attacked only bacteria that were resistant to the first phage. The dynamic duo did the job. Smith could cure newborn calves with a single dose even if they didn't get any colostrum, which contains antibodies from the mother cow and is essential for the health of the newborns.[15]

Félix d'Herelle would have got a real kick out of Smith's trials, not just because they were so successful, but because Smith was able to confirm what d'Herelle had already observed in the earliest experiments of phage therapy. The mere contact with a treated animal could suffice to immunize an untreated animal. Like d'Herelle's chickens that were protected from fowl typhoid as soon as they pecked in the phage-laced droppings, Smith's calves did not contract diarrhoea if they were put in an uncleaned pen where another young cow had been cured of diarrhoea using phages. If Smith sprayed the pens with small amounts of phages, he saw the same prophylactic effect. The curative properties were contagious, just as d'Herelle had proclaimed in the past.

Before Smith's very eyes another piece of early history unfolded in the experimental barn. D'Herelle had claimed that naturally occurring phages could take care of the healing of infectious diseases like dysentery or cholera. Hardly any other researcher – no matter how much of a phage believer he or she might be – would have agreed with his daring claim. However, one day an unknown phage turned up in Smith's barn that protected the calves from diarrhoea. In some calves' intestines, it fought the bacteria better than the viruses Smith had used.[16]

The fact that d'Herelle's theory is apparently valid in certain cases and that naturally occurring phages sometimes determine whether a patient will be healthy or not was recently demonstrated by a student of phage researcher Elizabeth Kutter. Peter Varey set out to learn more about the bugs of the 'hamburger disease' E. coli O157:H7. To do this, he left Kutter's lab at Evergreen College in the state of Washington and worked for a while at an institute of the US Department of Agriculture in Texas. There researchers infected cows or sheep that did not have the O157 bugs with O157 on a regular basis in order to find ways to rid the animals' intestines of the microbe. However, one flock of sheep couldn't be infected with O157. Their faeces kept turning out to be free of the microbes despite the fact that they hadn't been treated. The researchers couldn't explain this, but the phenomenon reminded Varey of d'Herelle's claim. The student examined the sheep droppings and isolated a phage that destroyed O157 in the test tube and protected the sheep from being colonized by the bug.[17]

Smith had done research on his rediscovery for over six years. Now, in 1986, he went into retirement. The phages had worked so well in his hands that he would have liked to continue doing research on the therapy. After all, there were still a few problems to be solved. Many infections are caused by several *E. coli* strains in life outside the confines of the lab. If Smith simulated this in experiments, he could cure some cows with phage mixtures, but in others resistant bacteria turned up that were still very dangerous. Smith conjectured that the different *E. coli* strains in the intestine exchanged phage resistance and virulence genes among themselves, and this was how new dangerous variants emerged.

In his last publication, Smith indicated that he had already found new phages that managed to deal with these variants.[18] However, after he retired he was unable to find anyone to fund his research. Willie Smith's last fantastic idea – phage therapy – was over 60 years old when he came up with it in 1980, but in the West it was still ahead of its time. The pressure on the agricultural industry to halt the excess of antibiotics in farmyards and look for alternatives was not yet great enough. Soon after, in the summer of 1987, Smith died.

## A phage shower for newly hatched chicks

A decade later, his seed germinated. Smith's experiments laid the foundation for today's development of phage therapies for animals. Two companies that have made the most progress are the Canadian branch of GangaGen and the US company Intralytix. GangaGen is developing a phage drug that is supposed to remove the O157 hamburger bug from cows before they are slaughtered. After successful experiments on some cows, GangaGen has been granted approval by the Canadian Food Inspection Agency to carry out large-scale trials.[19] Intralytix, which is based in Baltimore, Maryland is concentrating on *Salmonella* and *Listeria*. While *Salmonella* like to infest chickens, they don't make them sick. *Listeria* are found on meat, fruit, vegetables and certain types of cheese.

Battery egg farms are breeding grounds for *Salmonella* because of the cramped conditions in the henhouses. The shells of freshly laid eggs are often infested with these bacteria. However, they cannot be washed because the chicks wouldn't be able to hatch. Their stay in the incubator not only creates optimal conditions for the chicken embryos to develop, but promotes the growth of germs on the shell as well. When the chicks hatch, the germs on the shell rub off on them. Although they are washed off by chlorine bleach, 25 per cent of chickens sold in stores are contaminated with *Salmonella*.[20]

The Intralytix researchers have found a few stages in the industrial life of fattening chickens during which they can use phages to reduce the danger of *Salmonella*. When they spray freshly laid eggs with their cocktail of viruses from the water of Baltimore's harbour, they can lessen the number of *Salmonella* on the eggs 1000 times over. After the chicks hatch, they are moved to their new location by conveyer belt under nozzles that spray them with vaccines. If phages are added to the shower, most of the chicks remain free of *Salmonella*. Intralytix researchers also tested the possibility of washing the slaughtered fowl with phages and were able to reduce the *Salmonella* load of the chickens by 96 per cent.

The US Environmental Protection Agency (EPA) granted Intralytix permission to test its *Salmonella* phages outside the lab in slaughterhouses. A drug called LMP-102, which sanitizes *Listeria* from the surface of frankfurters and other meat products, is even being examined by the FDA for approval. If *Listeria* contamination could be lowered for all affected deli meats to one-tenth of the current level, the number of elderly people who die from this cause could be cut by half, an FDA report estimates. 'Our experiments show that we can use LMP-

102 to achieve this kind of reduction', says Alexander Sulakvelidze of Intralytix. 'If we can reproduce this in real life, I expect public health implications would be significant.'[21] Soon American consumers will be able to buy chicken breasts and sausages that phages have kept free of *Listeria* – and antibiotics to boot. For Sulakvelidze, a Georgian living in exile and co-founder of Intralytix, using phages in animal husbandry and food production has two advantages: 'First of all, perhaps we can someday stop using antibiotics completely and reserve their use for humans. Secondly, phages are probably among the most environmentally friendly drugs for meat production.'

Lee Jackson's phages are another eco-friendly alternative. They help American farmers to keep their tomatoes from contracting bacterial spot, which is extremely destructive. Jackson's phage cocktail, called AgriPhage, is more effective than the copper treatments or antibiotics conventionally used in the US and to which many bacteria have become resistant – and without damaging the beneficial microbes in the way antibiotics do.[22]

Jackson is actually ahead of Intralytix and GangaGen. Since 2002, phages have been available to American farmers, making AgriPhage the first phage therapy product approved in the US since penicillin was introduced. According to Jackson, demand was great from the start: 'People keep calling who want phages or who want us to develop phages for fire blight or other plant diseases. Everyone is urging us to hurry up.' Jackson's mini-company Agriphi has morphed into Omni-lytics. While the Salt Lake City-based enterprise is not sharing its sales figures, its success appears to be so explosive that it can afford to reverse the trend that recently turned many phage start-ups from the demanding field of human therapy to agricultural uses. In addition to developing phages for infected tomatoes, Omnilytics is now also looking for ways to help sick people.[23] This confirms the prognosis of phage researchers Bruce Levin and James Bull: 'Phage success in these agricultural endeavours will be a stepping stone for their development for human medicine', they wrote in 2004 in an article in *Nature Reviews Microbiology*.[24]

## 'Why didn't somebody do something?'

In order for most phage companies to survive, the detour via veterinary medicine is inevitable. Asher Wilf, whose Phage Biotech company had to do an about-turn from human medicine, sees large companies increasingly express interest in his phages, which are designed to cure infections in shrimp: 'Suddenly Fortune 500 companies from the agricultural sector have started calling us', he says. 'In a few months we're going to manage our breakthrough.' Yet that's no help for the patients who call him every day asking for phage drugs. They need a financially strong company that is prepared to set aside some money for researching phage therapy in humans. So far no money of this sort has appeared. On the contrary, most pharmaceutical companies are tending to pull out of the development of antibiotics.

Now something has changed, though. Just recently a multinational has started investigating the potential of phage therapy, clearly taking everyone by surprise. It's Switzerland's Nestlé food company. Until recently, probably no Nestlé manager would have dreamt that he or she would be leading the company in this particular adventure. The man behind this surprising beginning isn't a manager, but microbiologist Harald Brüssow. He is just the kind of researcher a researcher should be. Brüssow is so spellbound by his work that he has turned down several promotions in order not to exchange his lab bench for a desk. At the Nestlé Research Centre, the died-in-the-wool researcher is in the right place. The institute sits up on a green hill high over Lake Geneva. There, 300 scientists are on the quest for new products that may make money for the company some day. However, many of them are allowed to spend some time working on projects that seem so esoteric that it's hard to imagine that a new dessert or cheese spread may emerge from them in the future.

Actually, it was Brüssow's task to fight phages. The tiny viruses often get in Nestlé's way. Everywhere that milk is fermented, for instance in the production of yoghurt or cheese, the phages can strike and massacre the fermentation bacteria. 'Every day our production people are plagued by the fear that the phages will prevent them from delivering their product', says Brüssow. This isn't just a minor problem for the head of a factory that makes 500,000 litres of milk into mozzarella cheese every day. For this reason, Brüssow studied the battle between the bacterium *Streptococcus thermophilus*, which coagulates milk into mozzarella, and its natural enemies, the phages. After a while he had a plan: he wanted to genetically modify a bacterium so that it would become resistant to phages. But Brüssow forgot to keep Nestlé's customers in mind as he was designing his plan. The European population is so sceptical about genetic engineering that he and his bosses shelved the project.

Considering all the know-how that had evolved during the project, this was a real shame. Brüssow's superiors asked him if he had an idea for using the knowledge he had acquired in some way. And he did. After reading the old phage therapy studies and talking to his brother-in-law, he came to the conclusion that it was time to turn the tables. From then on, Brüssow broke rank with the bacteria and started fighting for the cause of phages. At that point, neither he nor his bosses had any idea where this journey would lead and what Nestlé, a food producer, would do with a phage drug if their experiments were successful.

Brüssow selected *E. coli* diarrhoea in children as the disease to target. He already had experience with these intestinal bacteria when he studied another diarrhoea therapy – unsuccessfully. 'So I still had something to settle with the *E. colis*.' Bacterial diarrhoea of this kind is rare in industrialized countries, but in the developing world it causes thousands of young children to become sick and die every year. Antibiotics do not help them to recover, even if the bacteria aren't resistant per se.

Within a short time, Brüssow painstakingly carried out preliminary studies that are among the best trials ever performed in phage therapy. Along with his colleague Sandra Chibani-Chennoufi, he investigated the behaviour of *E. coli* bugs and phages in mice.[25] This was necessary because little was known about either of them, although the harmless strains of *E. coli* bacteria are among the most important residents of the human intestine. During their trials, the two researchers stumbled upon an interesting phenomenon: when they fed mice with phages, those *E. coli* bacteria that permanently reside in the intestine are not affected. But the phages can attack and control the *E. coli* strains that the mice ingested before the trial.

Brüssow and Chibani explained this difference as follows: the resident *E. coli* bacteria sit deep enough in the mucous membrane to be protected from the phage attack. However, at least in mice, the penetrating *E. coli* germs don't appear to colonize the intestine such that they can avoid the viruses. 'Now we need to test whether this works the same way in people suffering from diarrhoea', says Brüssow. He has already carried out an initial small-scale safety test on 15 fellow researchers. Brüssow didn't have any trouble finding volunteers at the institute. 'I would have participated myself', he says, 'but as the head of the study I wasn't allowed to.' The test confirmed that the ingested *E. coli* phages had no side effects and left the subjects' healthy intestinal flora alone.[26]

Brüssow would like to start with treatment tests in children in Bangladesh soon. The company management has given the go-ahead for the clinical studies. 'Nestlé can definitely invest more money than small companies can', says Brüssow. He isn't yet able to ensure that the diarrhoeal bacteria don't entrench themselves so deep in the intestine that phages are powerless to reach them. Brüssow wants to investigate this in more detail in mice before he starts the treatment tests, but doing this in the intestine isn't exactly child's play. 'If I didn't have any constraints, I would have examined phage therapy in wound infections', he says, 'because it's easier to observe the processes there.' But Brüssow doesn't have free rein when it comes to his research. Biological drugs for diarrhoea have a plausible link to a company that sells probiotic yoghurt, but wound treatment? Maybe Nestlé will add the phages to its 'Allhydrat' preparation some day. The mixture of salts and glucose helps children to recover from severe diarrhoea. Whether the developing countries where Allhydrat is used could even afford it is open to speculation. Perhaps Nestlé will subsidize the phage drug because it will be good for its image. Nothing has been decided yet, and Brüssow will only say: 'It will be difficult to solve all these problems. But for me, the end product isn't what's most important. I want to critically test the possibilities of phage therapy. If a point comes when we are defenceless against resistant bacteria, everyone will ask, "Why didn't somebody do something?"'

## Two naive questions with consequences

Harald Brüssow, Willie Smith and Maximilian Pichlmaier are all examples of free-thinking scientists who advance the cause of phage therapy, despite all the obstacles. Another pioneer is Carl Merril, who shares many characteristics of Willie Smith, the scientist who rediscovered phage therapy in the West. Merril, who is now 68, also began working with phages at a time when it could be a career stopper. The nonconformism of the researcher is easy to spot when you accompany him to the parking garage of the National Institute of Health. Merril's Toyota Prius sticks out among the cars of his colleagues. It was the first hybrid model and Merril bought it before most drivers were even aware that such a thing as hybrid technology existed. The researcher prefers to leave his two Ferraris at home. 'I'd be embarrassed to drive them here', he says. He's wearing a long brown jacket, despite the hot and humid mid-Atlantic weather. A wide-brimmed hat completes the Western image.

'In 1965 I was a junior scientist. I attended a course in molecular biology at the Cold Spring Harbor lab on Long Island', Merril relates. It wasn't just any old course, but a series of courses that had been established 20 years before by Max Delbrück, co-founder of the Phage Group. Its members studied the basic biology of phages in order to use the simple being to find out how life emerges from a conglomerate of molecules. As described in Chapter 4, the Phage Group created the foundation of molecular biology. Its members had no taste for phage therapy.

'I asked the course instructors two questions', Merril recalls. 'The first one was: aren't there any phages that infect human cells?' The reply was an indignant one: 'Of course not. They aren't called *bacterio*phages for nothing.' The great authorities' answer wasn't enough for Merril. In 1971, he carried out complicated experiments and proved that in rare cases, phages permanently smuggled DNA into human cells. This was a sensation at the time.[27] Genetic engineering had not yet been heard of, and no one had managed to insert a piece of hereditary material into human cells. Interestingly, only a few researchers have undertaken to study this phenomenon since then. However, Merril considers it to be so rare that it is not a factor that needs to be considered when assessing the safety of phage therapy.

Merril's second question at the course in 1965 was: 'Why doesn't anyone do phage therapy? Back then I had no idea that it in fact already had a long history.' And it seemed to be a naive question after Gunther Stent, a member of the dominating Phage Group, had pronounced his scathing verdict on the method only two years earlier in his textbook. Merril wasn't able to let go of the question, however. 'But at the beginning, I didn't think about it a lot because it didn't seem to make any sense. After all, we had antibiotics.' Despite this, in 1973, Merril did a few experiments in which he observed what happened with a certain type of phage when he injected it into mice. The result was that the phages quickly disappeared from the blood and the organs and landed in the spleen, whose task includes filtering foreign material out of the blood.[28] This finding did not speak for the effectiveness of phage therapy, since the phages that had been collected couldn't hunt down bacteria – except in the spleen.

'Twenty years later I read a book about the brewing resistance crisis. One morning when I stood under the warm water from the shower massage that my wife had just bought I came up with the idea of how we could find phages that aren't filtered out by the spleen so quickly.' Merril, who was working at the US National Institute of Mental Health, interrupted the experiments he was working on and, to the astonishment of some of his fellow researchers, began doing phage experiments.

He injected the same type of phages that he had used in 1973 into the bloodstream of some mice. After seven hours, most of the viruses had landed in the spleen, but there were still a few left in the blood that the spleen had missed. This is what Merril had his sights set on. He removed these rare viruses from the blood of the mice and amplified them in bacteria. Then he re-injected them into mice and started the cycle all over again, repeating this eight times. In the end, Merril was able to select out a phage he called Argo from the rest of the crew. It was a phage that had the ability to survive significantly longer in the blood. Now the question arose as to whether Argo would also work better in therapy.

With these trials, Merril was one of the first scientists to not merely isolate phages from nature for therapeutic use but also improve them at the same time. At about the time that Argo was proving its endurance, Merril made the acquaintance of Richard Carlton. He was fascinated by Merril's trick. Carlton founded Exponential Biotherapies and worked with Merril to develop a phage drug. Carlton recalls the first trials with Argo in detail:

Out of 100,000 normal phages we injected only a single one remained in the blood after 18 hours. That isn't a good drug. In the case of Argo phages, out of 100,000 of them, 63,000 were still there after the same period had elapsed. Instead of being filtered out of the blood after two minutes, they stayed in the bloodstream for 24 hours. And that makes the difference. When we injected mice with bacteria and normal phages they became really sick, more or less equivalent to the shape patients are in when they are in intensive care. But the mice that were injected with bacteria and Argo hardly got sick at all. It took a trained eye to see that these mice were a bit less active.[29]

Not every phage is filtered out of the blood by the spleen as busily as the one Merril began his experiment with. There are phages that remain in the bloodstream long enough for healing to take place even without the selection trick. In his first experiments on mice, Williams Smith had observed the phages patrolling the blood for a much longer period of time. But Merril's experiment showed that natural phages could definitely be improved. In November 2002, he was able to report on his progress in a forum that he considered appropriate – the Cold Spring Harbor lab on Long Island, the same place he had stuck out with his impertinent questions. Other researchers are now picking up Merril's biotechnological trail.

While Carl Merril retired in autumn 2005, he is continuing to work on phage therapy with Sankar Adhya of the NIH. The two researchers cannot complain that they have a lack of ideas for improving phages with the help of genetic engineering tools. For one, they are considering how they can expand the narrow host range of phages. That would be extremely useful because it could allow the FDA's reservations regarding phage cocktails to be taken into account.

For the two researchers, a phage is serving as a model and launching pad that evolution itself has equipped with a wider appetite. Like Smith's virus, it uses its tail fibres to dock onto the K1 molecule on the surface of the *E. coli* strain bearing the same name. In addition, it has a second tail fibre protein with which it can bind to the surface molecules called K5 of the *E. coli* strain K5. In this way a phage can infect two bacterial strains. The genes for the two docking molecules lie directly behind each other in the genome. If Merril and Adhya insert genes for additional docking molecules, they could expand the phage's appetite even further.[30]

## A single phage molecule beats a whole microbe phalanx

The research group led by James Norris of the Medical University of South Carolina is carrying out a fundamental reconstruction in its phages that goes far beyond genetically expanding the menu. The phages are modified so radically that they cease to be functioning viruses. Instead, they're not much more than a phage shell – a head, a tail and tail fibres – that serve as emissaries carrying lethal messages to the targeted bacteria. These messages are genes that are translated in the targeted bacterium into toxic proteins. Due to the quick death of the host bacillus and because the researchers have inactivated some required genes from the phage, it cannot multiply except in very special laboratory conditions. In so doing, Norris is dispensing with an advantage of normal phage therapy: the constant multiplication of the effective agent at the site of the infection. But he is also avoiding a potential danger. As described in Chapter 6, under certain circumstances, phages can distribute toxins that are dangerous for humans among the bacteria. With Norris's impotent, genetically engineered phages, this distribution mode is halted.[31]

Other research groups and companies are pursuing similar projects. They intend to change the phages so that as many unknown and, in turn, potentially dangerous parts as possible disappear. This strategy could have considerable appeal with the public. A complex organism that many patients have mixed feelings about because it is a virus is turned into a simpler, 'cleaner' drug. One scientist was so successful with this strategy that he managed to reduce a bacteriophage consisting of dozens of proteins and genes into one single protein – and a powerful one at that.

When a phage has entered a bacterium and produced offspring, it is confronted with the problem of how to leave its victim. Many phages possess a powerful protein duo that clear a path through the bacterial wall to the exterior. A protein called holin forms a channel through the cell membrane that surrounds the bacterium as an inner shell. The second protein, lysin, moves out through these gaps and gnaws away the connective struts of the stable wall surrounding the outside of the bacterium. The microbes burst and the young phages stream out looking for new prey.

More than 30 years ago, Vincent Fischetti had used lysin as a tool in the lab without realizing the potential that was slumbering in his sample vials. 'Back then I was interested in some proteins from the cell wall of the streptococci and wanted to purify them', Vince Fischetti recalls, now a professor at Rockefeller University in New York City. 'I used lysin to purify the proteins I was interested in.' For Gram-positive bacteria like *Staphylococcus* and *Streptococcus*, from the exterior no cell membrane is in the way, so not even holin is needed to open them up; lysin does the job all by itself.

When Fischetti sought a solution for the growing antibiotic resistance crisis a few years ago, he recalled his old experiments. Couldn't lysin be used to treat infections with Gram-positive bugs? Fischetti was so captivated by the idea that he started experimenting right away. The phage component lysin shares a great advantage with the entire phage: its specificity.

Every phage that attacks Gram-positive bacteria produces its own lysin, which is optimally tailored to the victim bacterium. The enzyme does not attack other bugs. Fischetti's group was able to show that lysin from *Streptococcus pyogenes* phages only destroys *S. pyogenes* bacteria but spares closely related streptococci.

Much more exciting, however, was the efficiency with which the lysin scarfed up the bacteria. Fischetti has stored a video on his computer: ten billionths of a gram of lysin are dripping from a pipette into a culture with ten million streptococci. Within five seconds, the cloudy soup becomes completely clear. 'It's sterile', says Fischetti. 'And it also works with antibiotic-resistant streptococci.' You can see that even after showing the video countless times, he is still satisfied by it. In another trial, Fischetti tested the lysin on an animal – 2.5 billionths of a gram were enough to clear the oral cavity of a mouse of 10 million streptococci that had been administered to it shortly before. Soon afterwards, Jutta Loeffler of Fischetti's team managed the same thing with *Streptococcus pneumoniae* and the matching lysin.[32]

Both experiments open up interesting perspectives. *S. pyogenes* is the agent of strep throat, which primarily strikes children and can have serious long-term effects such as endo-carditis, an inflammation of the lining of the heart, and rheumatic fever. In daycare centres, sometimes up to 50 per cent of the children harbour *S. pyogenes* in their throats, although they don't exhibit any symptoms. These carriers can spread strep throat.[33] If they were treated with lysin, this method of transmission could be halted – a process that antibiotics can't be used for because they fuel resistances.

The case of *S. pneumoniae*, the agent of pneumonia and ear infections, is similar. With *S. pneumoniae*, the victims are also primarily children and elderly and weak people. In the US alone, these bacteria cause approximately 60,000 cases of pneumonia every year, and 10 per cent of them end in death. *S. pneumoniae* also resides in numerous noses and throats and could be driven away using lysin – without side effects and without damaging closely related streptococci that live there and protect people from infections. Lysins seem to have another big advantage: despite extensive investigations, Fischetti's team hasn't yet been able to track down any resistant bacteria. They exposed bacteria to increasingly higher concentrations of lysins and weren't able to find a single resistant cell. When they confronted the bacteria with antibiotics instead of lysins in analogous tests, they discovered many resistant cells. Fischetti explains this as follows: for the phages, it is essential not to get stuck inside their victims. For millions of years, lysins were selected in such a way that they attack the cell wall at positions that cannot be changed without lethal consequences for the bacteria.

Here's a hypothetical example: a lysin attacks the cell wall at a point at which five types of strut molecules come together. If a strut type were so mutated that the lysin could no longer grab it and cut it up, the bacterium would be resistant. Due to the same mutation, however, the other four strut molecules could also no longer attach to the modified molecule. This would cause the bacterial wall to be too unstable, and the bacterium would no longer be able to live. Only the simultaneous occurrence of mutations to all five types of struts, so that the struts fit together again, would make the bacterium resistant and establish a stable cell wall. But the probability of this happening is too small. We humans can now harvest the fruit of this long battle between phages and bacteria.

'Some companies are especially interested in *S. pneumoniae* lysins. Clinical studies are ready to go', says Fischetti. 'It all comes down to timing. If we had done these experiments ten years ago, no one would have been interested in them.' In the case of the third lysin Fischetti examined, he proved his sense of timing – which in this case turned out to be macabre. The events of 11 September 2001 hadn't yet transpired when he set his sights on a new target: *Bacillus anthracis*. For insiders, the anthrax agent was one of the most dangerous biological weapons even before it was used in the anthrax attacks in the US in autumn 2001. The bacilli, which claimed the lives of five people back then, develop their efficacy insidiously. Ground to extremely fine dust, they are inhaled as spores by the unsuspecting victim. In the lungs, the macrophages, cells of the immune system, transport them into the neighbouring lymphatic vessels. There the spores turn into growing bacteria that produce toxins. After between two and five days, the symptoms of a regular flu occur: fever, headache, coughing, nausea and weakness. The toxins attack the macrophages and make the affected lymph nodes melt. Their tissue dies and blood flows into the body cavities between and around the lungs and the heart.

Suddenly, the violent battle between the body and the bacteria also becomes clearly visible on the exterior: the fever skyrockets and blood pressure drops. A whistling sound announces the victim's shortness of breath because the dying, bloated lymph nodes make the windpipe cave in. The only thing that can help is artificial respiration. The bacteria advance into the blood and spread their toxins there, causing the immune system to overreact and set off a shock. Up to a litre of fluid and more trickles into the abdominal cavity and the crack between the lining of the lungs and the chest. The heart begins to race and soon it switches to a slow and laborious rhythm until it completely stops – often barely half a day after the victim has been admitted to the hospital as a result of the high fever and low blood pressure.

The experts' uneasiness has increased even more since rumours started spreading that artificially manufactured antibiotic-resistant anthrax bacilli are being stored in Russian labs. There are also supposed to be variants against which the US military's vaccine is ineffective. For this reason it wasn't surprising that the US Department of Defence was extremely interested in Fischetti's new goal and financed the project. With this tailwind, his people quickly isolated a lysin from an anthrax phage and tested it. They injected mice with a million anthrax bacilli in their abdominal cavity. After a maximum of five hours, they were all dead. The site of injection showed a huge oedema and blood poured out of their eyes and mouth. When the researchers injected 50 millionths of a gram of lysin 15 minutes after the lethal dose of bacilli, they could save 13 of the 19 mice.

In August 2002, Fishetti's team reported its triumph in the prestigious scientific journal *Nature*.[34] 'Since then everything is different', says Fischetti. 'I get countless calls from pharmaceutical companies. My whole career has changed. Now everything is focused on lysin.'

A rushed programme is intended to bring the drug to hospitals as quickly as possible. Now that the tests on mice have been conducted, the phage lysin also has to prove itself in anthrax-infected monkeys. In the end, a study will follow that is intended to confirm its harmlessness for humans. Therapeutic trials on humans are not possible because the disease is dangerous and rare. When the programme has been concluded, 'the military will store masses of the anthrax lysin', says Fischetti. In his lab, new experiments with lysins against multi-resistant *Staphylococcus aureus* and enterococci are underway.[35]

Humanity is fighting back – with the help of bacteriophages.

# 8

## what's the future of phage therapy?

More than 80 years have passed since Saturday, 2 August 1919, when Félix d'Herelle had Robert K, a patient with severe dysentery, swallow bacteriophages, ringing in the era of phage therapy. Robert's bloody diarrhoea stopped that same evening. Soon afterwards the phage medicine cured four other children. Despite this, d'Herelle warned that this was not absolute proof that phages were effective.

When he wrote his memoirs 25 years later while under house arrest, d'Herelle was convinced of the efficacy of phages. In his view, countless experiments had provided hundreds of instances of evidence that phages can heal. They not only cured dysentery, but a whole series of bacterial infections like plague, cholera and typhoid fever. The articles he wrote before his death in 1949 attest to his unshakable belief in the healing power of phages.[1] The contradictory studies carried out by other researchers and the headlines lauding penicillin, which had been available from 1944, couldn't change his conviction that phages were the true miracle drug.

Today it's clear, however, that neither phages nor antibiotics are the almighty panacea. While antibiotics continue to be among the most significant drugs, bacteria's resilience keeps demanding new weapons.

One of these weapons could be phages. But what do potential scenarios for their use look like? Phages will probably be used most in cases where antibiotics fail, as Frenchman Jean-François Vieu demonstrated between 1950 and 1980. In an article in *Nature Reviews Drug Discovery*, phage researcher Carl Merril wrote that luckily bacteriophages are likely to be best suited for use in such cases.[2] This is because studies have shown that the resistant bacteria of one species are often closely related worldwide. For example, a team led by Alexander Tomasz of Rockefeller University in New York City found that 70 per cent of 3000 multi-resistant *Staphylococcus aureus* bacteria worldwide belonged to only five strains.[3] Consequently, it could suffice to find a few phages that attack these five global strains in order to make a drug available that is active against three-quarters of infections with resistant staph.

However, phages cannot be used equally well for all types of infections. Their impressive size compared to chemical molecules and the human body's tendency to degrade and excrete them, more or less quickly depending on the type of phage, force researchers to look for suitable areas of use. All the superficial infections that can be easily reached by phages are promising: infected wounds in the case of diabetes, after operations or in the case of burns – the focus of Nodar Danelia and the Polish phage therapists from Wroclaw. Burn wounds are often colonized by multi-resistant bacteria such as *Pseudomonas aeruginosa*. Then there is always the danger that the bugs will advance into the blood. In addition, doctors cannot transplant skin. 'Over 60 per cent of the patients who die of infected burns become victims of *Pseudomonas*', says Sergey Bujanover of the Israeli company PhageBiotech.[4]

A question that remains open is what model of phage use could catch on: the one the FDA requires, with its demand for single-phage drugs, or the customized phage cocktail, which pioneers like Félix d'Herelle and Georgiy Eliava developed and Nodar Danelia would like to use in Germany.

In the case of the single-phage model, the phage drug is treated like any other drug. Let's suppose that Richard Honour's company Viridax were to gain approval for its staph phage. It would sell the drug all over the world, like Pfizer sells the antibiotic Zyvox for MRSA or Bayer sells Cipro for anthrax and other bacteria. The success of this model will primarily depend on whether the firms actually find phages that cover enough strains of a bacterial species for day-to-day use in hospitals – and how long it will take for resistant bacteria to emerge.

The system that Danelia wants to establish is based on a large phage bank from which the matching phage is selected once the exact diagnosis has been made. However, it may not be financially viable. How expensive is it to build up a phage bank that is large enough and get it approved? Is it affordable to select or modify the correct phage mixture for each patient?

Much work is still required in order for either model – or both models – to become established. After several successful animal trials, phage researchers now have to produce clear evidence that phages can also cure people. A number of experts expect that this will be possible: 'On the basis of the results so far, there is good reason to believe that the development of phage for treating and preventing bacterial diseases will be successful, at least in limited settings', wrote Bruce Levin and James Bull in *Nature Reviews Microbiology* in 2004.[5] The verdict of these two researchers holds even more weight because up to now they have not been known as phage therapy enthusiasts.

In addition, phage therapists have to convince regulatory authorities and doctors that their drugs are safe. This will require a great deal of persuasive effort on their part. Experiments performed by Hans Ochs of Washington University in Seattle may provide some assistance. For over three decades, he has been injecting phages into the bloodstream of human patients in order to study the immune system, and he has not observed any toxic effects.[6]

In order to do these studies, more money is required than has been available so far. While some government funding is now being channelled into phage therapy research, most of it benefits veterinary medicine. Many start-ups have also turned away from human medicine and now focus on animal medicine. A ray of hope has been provided by Nestlé's involvement in investigating phage therapy for *E. coli* diarrhoea. Still, compared to the funding that is normally pumped into drugs until they are ready for use, investments in phage therapy have been minimal.

Yet time is running out. Thousands of people are suffering from infections with antibiotic-resistant bacteria, which have been responsible for many of them losing a limb or even their lives. Every time phage researcher Elizabeth Kutter is quoted in a newspaper article or TV report on phage therapy, she gets calls from victims of bugs asking her for help. One of them was Saharra Bledsoe, the sister of diabetic Fred Bledsoe. He stepped on a nail in April 2002, and the wound became infected. Six months later, Bledsoe's doctors had scheduled an appointment to amputate his foot, when US broadcasting company CBS showed a report featuring Kutter. Like musician Alfred Gertler and other patients from the West, Fred Bledsoe decided not to have his foot amputated and chose to go to Georgia instead. Georgiy Eliava's progeny used phages to treat him and saved a Western patient's foot yet again.[7]

These stories sound good: patients who cannot be helped by anyone in the West end up finding help in the East. Yet in order for all humankind to win the battle against multi-resistant bacteria, it will take more than a trip to Georgia.

# appendix 1

## a short list of bacteria[1]

| Bacterium | Disease[2] | Remarks | Phage therapy | Page |
|-----------|-----------|---------|---------------|------|
| *Acinetobacter baumannii* | Opportunistic infections (lung) | Primarily infects patients with weakened immunity. Mortality 25–50 per cent (pneumonia). | – | 26, 47 |
| *Bacillus anthracis* | Anthrax (various forms: skin, lung) | One of the most important bacteria used in germ warfare (used in the anthrax attacks in the US after Sept. 11, 2001). Pulmonary anthrax has a high mortality. | A phage enzyme that dissolves anthrax bacteria in a highly specific manner has proved effective in animal experiments (see Chap. 7, note 34). | 245 |
| *Bordetella pertussis* | Whooping cough | Primarily infects children. A safe and effective vaccination is available. | – | 163 |
| *Campylobacter* spp.[3] | Diarrhoea, enteritis | Transmitted primarily through undercooked chicken. | Dutch researchers are working on reducing the bacterial load in *Campylobacter*-infected chickens by using phage therapy.[4] | 223, 224–5 |
| *Clostridium difficile* | Opportunistic infections (gastrointestinal) | Can colonize the alimentary canal of patients who have been treated with antibiotics. | In animal experiments *C. difficile* intestinal infections have been cured with phages (see Chapter 5, note 91). | 159 |

| Bacterium | Disease[2] | Remarks | Phage therapy | Page |
|---|---|---|---|---|
| Clostridium spp.[3] | Gas gangrene | Severe infection of wounds by anaerobic bacteria; frequent infection during war in earlier times. | In the Soviet Union during the Second World War. | 153 |
| Corynebacterium diphtheriae | Diphtheria | Childhood disease. A safe and effective vaccination is available. | – | 52, 198 |
| Escherichia coli | Gastrointestinal infections, urinary tract infections, pathogenic for animals or humans depending on the strain | Normal inhabitant of the human intestine, infects patients with weakened immunity. E. coli that are equipped with additional toxic or virulence genes also infect healthy humans. Particularly virulent strains like O157:H7 made the headlines as the culprit behind 'hamburger disease'. | In the early phase of phage therapy. Practised today in Russia and Georgia, clinical studies in Bangladesh soon (see Chapter 7). Veterinary experiments with cows and chickens in the UK, the US and Canada (see Chapter 7, notes 11, 15 and 16). | 43, 71, 92, 166, 168, 198, 223–5, 226, 230, 236, 241 |
| Haemophilus influenzae (different types) | Meningitis, among other diseases | Primarily infects children in a whole number of organs. There is a vaccination for H. influenzae type b (Hib). | – | 25 |
| Klebsiella pneumoniae | Opportunistic infections (lung) | Primarily infects patients with weakened immunity. Mortality 25–50 per cent (pneumonia). Often resistant to antibiotics. | In Georgia. | 25, 45, 215 |

| Bacterium | Disease[2] | Remarks | Phage therapy | Page |
|---|---|---|---|---|
| *Listeria monocytogenes* | Listeriosis | Infection through contaminated dairy products and vegetables. Insidious because the bacteria can also multiply in the refrigerator. | In Switzerland and the US, experiments are being conducted with a phage enzyme that dissolves listeria in a highly specific manner in order to keep cheese rinds free of the bacteria (see Chapter 7, note 32). | 224, 232 |
| *Mycobacterium tuberculosis* *M. bovis* *M. africanum* | Tuberculosis (various forms: tuberculosis of the lung is most frequent) | Increasing rapidly in African countries and Russia, among others. Multi-drug resistant (MDR) tuberculosis drastically increases the cost of treatment. | At least two scientists are doing research to develop phage therapy. The mycobacteria hide in the interior of human cells, making it extremely difficult for phages to target them (see Chapter 4, note 23). | 15 52 115 |
| *Neisseria gonorrhoeae* | Gonorrhoea (clap) | – | – | 16, 19 |
| *Proteus* spp.[3] | Opportunistic infections (lung, among others) | Primarily infects patients with weakened immunity. Mortality 25–50 per cent (pneumonia). | In Georgia. | 166, 168, 200, 206 |
| *Pseudomonas aeruginosa* | Wound infection (burns), pulmonary infection in the case of cystic fibrosis | Primarily infects patients with weakened immunity. Frequent cause of death in the case of severe burns and cystic fibrosis. | In Georgia and Russia (burns). | 26, 166, 195, 200, 204, 212, 215, 249 |

| Bacterium | Disease[2] | Remarks | Phage therapy | Page |
|---|---|---|---|---|
| *Salmonella enteritidis/ typhimurium* | Primarily gastroenteritis | Primarily transmitted by chicken and eggs. Multi-resistant strains like DT104 exist. | A company in the US is developing methods that use phages to reduce the *Salmonella* load of fattening hens. | 39, 214, 223, 231 |
| *S. typhi* | Typhoid fever | | In the US and Canada in the 1940s. | 70, 92, 101, 113, 229 |
| *S. paratyphi* | Paratyphoid fever | | In Germany in the 1940s (Behringwerke). | 92, 107 |
| *S. gallinarum* | Fowl typhoid | Aviary disease. | F. d'Herelle performed experiments in the 1920s. | 60 |
| *Serratia marcescens* | Opportunistic infections | Primarily infects patients with weakened immunity. Mortality 25–50 per cent (pneumonia). Often resistant to antibiotics. | – | 174 |
| *Shigella* spp.[3] | Bacterial dysentery (colitis with diarrhoea) | A whole series of species (*S. dysenteriae, S. sonnei, S. flexneri,* etc.) exist that are virulent to different extents. There is also a type of dysentery that is caused by amoebas. | In the 1920s, 30s and 40s throughout the world; afterwards in the Soviet Union; today in Georgia and Russia. | 48, 71, 73, 75, 92, 105, 150, 156, 162, 248 |

256

| Bacterium | Disease[2] | Remarks | Phage therapy | Page |
|---|---|---|---|---|
| Staphylococcus aureus | Wide range of types of infection, including sepsis, abscesses, furuncles, carbuncles, osteomyelitis | One of the most important germs transmitted in hospitals (referred to as nosocomial infections). Frequently multi-resistant. | In the 1920s, 30s and 40s in many countries; afterwards in the Soviet Union; today in Georgia, Poland and Russia, intensive research in the US and India. | Entire Chaps 1 and 2, 69, 74, 92, 100, 104, 117, 161, 168, 169, 200, 207, 214, 218 |
| Streptococcus spp.[3] S. pneumoniae (pneumococci) S. pyogenes Enterococcus spp.[3] | Wide range of types of infection Pneumonia Strep throat Opportunistic infections | — A vaccination is available. One of the most important germs transmitted in hospitals. Frequently multi-resistant. | Two phage enzymes that specifically dissolve S. pneumoniae and S. pyogenes respectively are effective in animal experiments (see Chapter 7, note 32). | 20, 200 244 244 Entire Chap. 2, 219 |
| Vibrio cholerae | Cholera | Many multi-resistant variants occur primarily in Africa and India. | In the 1920s, 30s and 40s in many countries, primarily in India and in the Soviet Union; afterwards experiments in the Soviet Union, WHO-sponsored experiments in Pakistan well into the 1960s.[5] | 82, 102, 129 |
| Xanthomonas campestris | Various infections from plants | — | A phage preparation for X. campestris infections in tomatoes is available for purchase in the US. | 233 |
| Yersinia pestis | Plague (pneumonic and bubonic) | — | In the 1930s. | 79, 102 |

# appendix 2

# the advantages and disadvantages
# of phage therapy

| | Advantage | Remarks | Page |
|---|---|---|---|
| 1 | Phages are very specific and do not harm the useful bacteria that live in and on the body. | As a result, there are no side effects like diarrhoea or secondary infections such as those that occur in treatment with antibiotics. See disadvantage 2. | 159 |
| 2 | Due to their specificity, phages do not cause a selection of resistances in the useful bacteria that live in and on the body. | | 159 |
| 3 | We are constantly ingesting phages. In general, they are harmless to human beings. When well-purified phages are used, few side effects have been described for all types of administration. | Because they are harmless, phages can be used for combating harmful bacteria in fattening animals and food. See disadvantage 8. | 54, 123, 221, 223 |
| 4 | Phages are an 'intelligent' drug. They multiply at the site of the infection until there are no more bacteria. Then they are excreted. | See disadvantage 5. | 111 |
| 5 | Bacteria that have become resistant to a certain type of phage continue to be destroyed by other types. | Bacteria that have become resistant to a certain antibiotic often become resistant to other drugs more easily. | 73, 162 |
| 6 | Phages are found throughout nature. This means that it is easy to find new phages when bacteria become resistant to them. | If each newly isolated phage requires approval, this procedure could become too expensive. | 162, 250 |

|   | Advantage | Remarks | Page |
|---|---|---|---|
| 7 | Evolution drives the rapid emergence of new phages that can destroy bacteria that have become resistant. This means that there should be an 'inexhaustible' supply. | The development of a new antibiotic for resistant bacteria takes several years. | 162 |
| 8 | Some resistant bacteria that have been selected during treatment with phages are less virulent and can be fought by the immune system. | Antibiotic-resistant bacteria are generally not less virulent. | 228 |
| 9 | Phages are also active against bacteria that have become resistant to antibiotics. | | 196, 208 |
| 10 | Phages can be genetically modified in order to make up for some of their disadvantages. | | 241 |
| 11 | Individual components of phages (e.g. lysins) can also be used as antibiotic substances. So far resistances have not occurred despite comprehensive testing. | | 242 |

|   | Disadvantage | Remarks | Page |
|---|---|---|---|
| 1 | There are no internationally recognized studies that prove the efficacy of phages in humans. | Numerous animal experiments demonstrate the efficacy against different infections. The first studies on humans are underway. | 209, 212, 215, 219, 226, 237 |
| 2 | The great specificity of phages is a disadvantage when the exact species of infecting bacteria is unknown or if there is a multiple infection. | For good results, the efficacy of phages against the infecting bacteria should be tested prior to application in the lab. For this reason, phages are less suitable for acute cases. Mixtures consisting of several phages can fight mixed infections. | 99, 168, 211 |
| 3 | Bacteria can also become resistant to phages. | See advantages 5, 6, 7 and 8. | 73, 162, 228 |

| | Disadvantage | Remarks | Page |
|---|---|---|---|
| 4 | Bacteria have a type of 'immune system' that destroys the hereditary material of some penetrating phages. Only suitable phages can conquer this 'immune system'. | The efficacy of phages needs to be tested in the lab prior to use in treatment. | 99, 121 |
| 5 | In comparison to chemical molecules, phages are relatively large. For this reason, the sites in the body that can be reached by them must be carefully clarified. | So far there have been too few pharmacological studies that have clarified these questions. Because the phages multiply as long as bacteria are present, in some cases it only takes a few phages in an inaccessible location in the body to bring about healing. It appears that phage therapy is best suited for infected sites such as wounds, where phages can be easily applied. | 73, 92, 249 |
| 6 | Infections whose agents are hidden in the interior of human cells may be inaccessible to phages. | In the 1940s, experiments were carried out that demonstrated good results for typhoid fever, an infection in which the agents seek refuge in human cells, at least to some extent. Researchers are trying to 'sneak' in phages by using genetic engineering. | 115 |
| 7 | Phages that are injected into the bloodstream are recognized by the human immune system. Some of them are quickly excreted and, after a certain period, antibodies against the phages are produced by the body. For this reason, it appears that one type of phage can only be used once for intravenous treatment. | Not all types of phages are quickly excreted. In addition, variants can be selected that can remain in the blood for a long time. The antibodies do not occur for one or two weeks. | 239 |

|   | Disadvantage | Remarks | Page |
|---|---|---|---|
| 8 | In comparison to chemical molecules, phages are complex organisms that can transfer toxin genes between bacteria. | The selection of strictly lytic phages, sequencing the hereditary material of phages and toxicity tests can minimize this type of risk. | 197 |
| 9 | The shelf life of phages varies and needs to be tested and monitored. | | 209 |
| 10 | Phages are more difficult to administer than antibiotics. A physician needs special training in order to correctly prescribe and use phages. | | 99, 211 |

# notes

## Chapter 1

1   Burke, J. P. (2003) 'Infection control – a problem for patient safety', *New England Journal of Medicine*, 348: 651–6.

2   Osborne, L. (2000) 'A Stalinist antibiotic alternative', *New York Times Magazine*, 6 March.

## Chapter 2

1   Quoted in web.mit.edu/popi/rubin.slides.pdf.

2   Small, I. (2000) in *Workshop Summary: Emerging Infectious Diseases – from the Global to the Local Perspective*, National Academy of Science (USA) p. 76f.

3   De Angelis, C. D. and Flanagin, A. (2005) 'Tuberculosis – a global problem requiring a global solution', *Journal of the American Medical Association*, **293**: 2793–4.

4   British Thoracic Society (2004) 'Tuberculosis rises in England and Wales by a fifth – whilst all other EU countries see falls in TB cases', press release, 24 March.

5   Nientit, C., (2002) 'Comeback einer Seuche', *Sonntagszeitung*, 12 May; *WHO Press Release 41*, 12 June 2000.

6   Statistics obtained from the Office for National Statistics (document: 'Number of death certificates with *Staphylococcus aureus* and MRSA mentioned as the underlying cause, England and Wales').

7   See Chapter 1, note 1.

8   *WHO Report on Infectious Diseases 2000*. Contains many of the resistance rates of various bacteria and data on resistance trends, diseases and agents mentioned in this chapter.

9   US figures: 1900–1940 tables ranked in the National Office of Vital Statistics, December 1947; English figures: Griffiths, C., and Brock, A. (2003) 'Twentieth century mortality trends in England and Wales', *Health Statistics Quarterly*, **18**: 5–18.

10  www.nobel.se/medicine/laureates/1945.
11  Rountree, P. M. and Thomson, E. F. (1949) 'Incidence of penicillin-resistant and streptomycin-resistant staphylococci in a hospital', *Lancet*, 17 October; Bertschinger, J. P. (1957) 'Le bactériophage', *Schweizerische Apothekenzeitung*, **95**: 479–87.
12  Stuart Levy is one of the most renowned experts on antibiotics. He has written a comprehensive popular science book on the topic of antibiotic resistance, *The Antibiotic Paradox*, Perseus Publishing, 2002.
13  From a lecture held at the 12th European Congress of Clinical Microbiology and Infectious Diseases, 21–24 April 2002, Milan.
14  Samore, M. et al. (2001) 'High rates of multiple antibiotic resistance in *Streptococcus pneumoniae* from healthy children living in isolated rural communities: Association with cephalosporin use and intrafamilial transmission', *Pediatrics*, **103**: 856–65.
15  From a lecture held at the 12th European Congress of Clinical Microbiology and Infectious Diseases in Milan (see note 13); MacFarlane, J. et al. (1997) 'Influence of patients' expectations on antibiotic management of acute lower respiratory tract illness in general practice: questionnaire study', *British Medical Journal*, **312**: 1211–14.
16  Millar, M. R. et al. (2001) 'Carriage of antibiotic-resistant bacteria by healthy children', *Journal of Antimicrobial Chemotherapy*, **47**: 605–10; Ainsworth, C. (2001) 'They're everywhere', *New Scientist*, 19 May.
17  Khairulddin, N. et al. (2004) 'Emergence of methicillin resistant *Staphylococcus aureus* (MRSA) bacteraemia among children in England and Wales, 1990-2001', *Archives of Disease in Childhood*, **89**: 378–9.
18  US figures: see Chapter 1, note 1; UK figures: 'Improving patient care by reducing the risk of hospital acquired infection: A progress report', National Audit Office, 2004. The Department of Health has commissioned a national survey on all infections in hospitals for the first time. It is expected to be finished in September 2006.
19  US figures: see Chapter 1, note 1 and 'National nosocomial infections surveillance (NNIS) systemic report, data summary from January 1992 to June 2004, issued October 2004', *American Journal of Infection Control*, **32**: 470–85; European figures are taken from Tiemersma, E. W. et al. (2004) 'Methicillin-resistant *Staphylococcus aureus* in Europe, 1999–2002', *Emerging Infectious Diseases*, **10**: 1627–34; 'The fourth year of regional and national analyses of the Department of Health's mandatory *Staphylococcus aureus* surveillance scheme in England: April 2001–March 2005', *CDR Weekly*, **15**(25): 1–4.
20  Pittet, D. and Widmer, A. (2001) 'Händehygiene: Neue Empfehlungen', Swiss-NOSO, 8: 1–11; 'Verschmutzte Endoskope', *Bild der Wissenschaft*, 5/2002, p. 26.

21    Beaumont, W. (2003) 'Caring for a loved one with MRSA', rapid response
      to Howard, A. J. et al. (2003) 'Mortality from methicillin resistant *Staphy-
      lococcus aureus*', *British Medical Journal*, **326**: 501a.
22    "Cleaner hospitals' – more important to patients than choice', press
      release, British Medical Association, 26 June 2005.
23    'Overseas op over superbug fear', BBC news (news.bbc.co.uk/1/hi/
      wales/3580189.stm).
24    'MRSA: how politicians are missing the point', *Lancet*, **365**: 1203.
25    Parliamentary Office of Science and Technology (2005) 'Infection control
      in healthcare settings', postnote 247.
26    Hiramatsu, K. (1997) 'Reduced susceptibility of *Staphylococcus aureus* to
      Vancomycin – Japan 1996', *Morbidity and Mortality Weekly Report*, **46**:
      624–6. The quotation is taken from the BBC documentary film 'The virus
      that cures'.
27    Sievert, D. M. (2002) 'Staphylococcus aureus resistant to vancomycin –
      US 2002', *Morbidity and Mortality Weekly Report*, **51**: 565–7; Weigel, L. M.
      et al. (2003) 'Genetic analysis of a high-level vancomycin-resistant isolate
      of *Staphylococcus aureus*', *Science*, **302**: 1569–71; Tabaqchali, S. (1997)
      'Vancomycin-resistant *Staphylococcus aureus*: apocalypse now?', *Lancet*,
      **350**: 1644.
28    Gonzales, R. et al. (2001) 'Infections due to vancomycin-resistant *Entero-
      coccus faecium* resistant to linezolid', *Lancet*, **357**: 1179; Tsiodras, S. et al.
      (2001) 'Linezolid resistance in a clinical isolate of *Staphylococcus aureus*',
      *Lancet*, **358**: 207–8.
29    'Methicillin-resistant *Staphylococcus aureus* infections among competitive
      sports participants – Colorado, Indiana, Pennsylvania, and Los Angeles
      County, 2000–2003', *Morbidity and Mortality Weekly Report*, **52**: 793–5;
      'Outbreaks of community-associated methicillin-resistant *Staphylococcus
      aureus* skin infections – Los Angeles County, California, 2002–2003',
      *Morbidity and Mortality Weekly Report*, **52**: 88.
30    Robinson, D. A. et al. (2005) 'Re-emergence of early pandemic *Staphylo-
      coccus aureus* as a community-acquired methicillin-resistant clone',
      *Lancet*, **365**: 1256–8.
31    Gillet, Y. et al. (2002) 'Association between *Staphylococcus aureus* strains
      carrying gene for Panton-Valentine leukocidin and highly lethal necro-
      tising pneumonia in young immunocompetent patients', *Lancet*, **359**:
      735–59; Dufour P. et al. (2002) 'Community-acquired methicillin-
      resistant *Staphylococcus aureus* infections in France: emergence of a single
      clone that produces Panton-Valentine leukocidin', *Clinical Infectious
      Diseases*, **35**: 819–24.
32    Miller, L. G. et al. (2005) 'Necrotizing fasciitis caused by community-asso-
      ciated methicillin-resistant *Staphylococus aureus* in Los Angeles', *New
      England Journal of Medicine*, **352**: 1445–53.

33   Teuber, M. (2002) 'Vom Segen zum Alptraum', *Unimagazin* (Zurich), **1**: 31–4.

34   *Report of the Joint Committee on the Use of Antibiotics in Animal Husbandry and Veterinary Medicine (Swann Committee)*, Her Majesty's Stationary Office, London, September 1969; Klare, I. et al. (1999) 'Decreased incidence of VanA-type vancomycin-resistant enterococci isolated from poultry meat and from fecal samples of humans in the community after discontinuation of avoparcin usage in animal husbandry', *Microbial Drug Resistance*, **5**: 45–52.

35   German statistics: *WHO Report on Infectious Diseases* 2000; numbers for England and Wales: '*Salmonella typhimurium* DT 104 infections in humans: monthly totals for 1999 to 2003', *CDR Weekly*, **14**(15): 6.

36   Molbak, K. et al. (1999) 'An outbreak of multidrug-resistant, quinolone-resistant *Salmonella enterica* serotype *typhimurium* DT104', *New England Journal of Medicine*, **341**: 1420–5; Ferber, D. (2000) 'Superbugs on the hoof?', *Science*, **288**: 792–4.

37   www.schweizerbauer.ch/news/aktuell/artikel/07965/artikel.html.

38   Davies, J. (1994) 'Inactivation of antibiotics and the dissemination of resistance genes', *Science*, **264**: 375–82.

39   Marshall, G. C. et al. (1997) 'D-Ala-D-Ala ligases from glycopeptic antibiotic-producing organisms are highly homologous to the enterococcal vancomycin-resistance ligases VanA and VanB', *Proceedings of the National Academy of Sciences of the United States of America*, **94**: 6480–3.

40   Waters, V. L. (2001) 'Conjugation between bacterial and mammalian cells', *Nature Genetics*, **29**: 375–6.

41   Brock, T. and Madigan, M. T. (1991) *Biology of Microorganisms*, Prentice Hall, 6th edn, pp. 257–61; Hughes V. M. and Datta, N. (1983) 'Conjugative plasmids in bacteria of the 'pre-antibiotic' era', *Nature*, **302**: 725–6.

42   Philippon, A. et al. (2002) 'Plasmid-determined AmpC-type, beta-lactamases', *Antimicrobial Agents and Chemotherapy*, **46**: 1–11.

43   Oliveira, D. C. et al. (2002) 'Secrets of success of a human pathogen: molecular evolution of pandemic clones of meticillin-resistant *Staphylococcus aureus*', *The Lancet Infectious Diseases*, **2**: 180–9.

44   Amyes, S. G. B. (2005) 'Treatment of staphylococcal infection', *British Medical Journal*, **330**: 976–7.

45   Leeb, M. (2004) 'A shot in the arm', *Nature*, **431**: 892–3.

46   Landman, D. et al. (2002) 'Citywide clonal outbreak of multiresistant *Acinetobacter baumannii* and *Pseudomonas aeruginosa* in Brooklyn, NY', *Archives of Internal Medicine*, **162**: 1515–20.

# Chapter 3

1   The memoirs *Les pérégrinations d'un bactériologiste* by Félix d'Herelle are
    located in the archives of the Pasteur Institute, Paris (Fonds F. d'Herelle).
    They were stored there along with other items from d'Herelle's estate
    (notes, diaries written by d'Herelle and his family, photographs, and so
    on) by d'Herelle's grandson Claude-Hubert Mazure. Mazure kindly
    provided the author with a complete copy. US science historian William
    C. Summers was the first person to evaluate d'Herelle's unpublished
    papers. He wrote the excellent scholarly biography *Félix d'Herelle and the
    Origins of Molecular Biology* (Yale University Press, 1999). The episode with
    the therapeutic trials is found in the memoirs (abbreviated as M in the
    following notes) on pages 404–6. See also d'Herelle, F. (1921) *Le bactério-
    phage – son rôle dans l'immunité*, Masson, pp. 211–14, and *The
    Bacteriophage – its Role in Immunity* (1922) Williams and Williams,
    pp. 266–71. In contrast to what is found in the memoirs, in the two
    books, d'Herelle reports that Robert K had 5–7 bowel movements per day
    as opposed to 12.
2   M376–7. The fact that the family helped with the vaccine production is
    also mentioned in d'Herelle's daughter Huberte's diary in the entry dated
    25 October 1917.
3   *Merck's 1899 Manual*, Merck & Co, facsimile from 1999.
4   Statistics from the German Federal Statistics Office.
5   US figures: '1900–1940 tables ranked in National Office of Vital Statistics,
    December 1947', US Census Bureau. English figures: 'Causes of death
    1900', Office for National Statistics, and Griffiths, C. and Brock, A. (2003)
    'Twentieth century mortality trends in England and Wales', *Health Statis-
    tics Quarterly*, **18**: 5–18.
6   Leven, K.-H. (1994) 'Die bakterielle Ruhr im deutschen Heer während des
    Krieges gegen die Sowjetunion, 1939–1945', p. 82, in *Medizin für den
    Staat – Medizin für den Krieg; Aspekte zwischen 1914 und 1945; gesam-
    melte Aufsätze* (eds by R. Winau and H. Müller-Dietz), Matthiesen.
7   D'Herelle, F. (1917) 'Sur un microbe invisible antagoniste des bacteries
    dysentériques', *Comptes rendus de l'Académie des Sciences*, **165**: 373–5,
    and M379.
8   Ruska, H. (1940) 'Über die Sichtbarmachung der bakteriophagen Lyse im
    Übermikroskop', *Naturwissenschaften*, **28**: 45–6. The publication
    appeared in 1940. Ruska made the first observations in 1939.
9   Basic information on the biology of phages can be found in publications
    such as Calendar, R. (ed.) (1988) *The Bacteriophages*, Plenum Press, and
    Madigan, M. T. et al. (2002) *Brock Biology of Microorganisms*, Prentice Hall.
10  Twort, F. W. (1915) 'An investigation on the nature of ultra-microscopic
    viruses', *Lancet*, **2**: 1241–3.

266 notes

11 Information on the experiments on fowl typhoid is available in M397–404, d'Herelle, F. (1921) Le bactériophage – son rôle dans l'immunité, Masson , pp. 150–62, and d'Herelle, F. (1919) 'Sur une épizootie de typhose aviaire', Comptes rendus de l'Académie des sciences, **169**: 817–19.

12 Based on a sentence coined by science historian Ton van Helvoort in 'Felix d'Herelle en de bacteriofaag-therapie: De laboratoriumtafel naast het ziekbed', Tijdschrift voor de Geschiedenis der Geneeskunde, Natuurwetenschappen, Wiskunde, en Techniek (1986) **9**: 118–31.

13 M409; in the French original, the buffalo epidemic was referred to as 'barbone'. It was a bacterial blood poisoning in cattle.

14 The information about d'Herelle's life from his birth until his departure from Canada to Guatemala is taken from M1–53. Some of the information about d'Herelle's parents and the bankruptcy of the chocolate factory was communicated by Mazure.

15 M54–228 for the time in Guatemala and Mexico.

16 M229.

17 D'Herelle, F. (1911) 'Sur une épizootie de nature bactérienne sévissant sur les sauterelles au Mexique', Comptes rendus de l'Académie des Sciences, **152**: 1413–15. Today the bacteria that d'Herelle described as Coccobacillus sauterelle are referred to as Aerobacter aerogenes (see Vincent, C. and Coderre, D. (1991) La lutte biologique, Gaëtan Morin).

18 D'Herelle, F. (1914) 'Le coccobacille des sauterelles', Annales de l'Institut Pasteur, **28**: 280–328 and **28**: 387–407. More information is found in M237–74.

19 M285.

20 Uvarov, B. P. (1928) Locusts and Grasshoppers, The Imperial Bureau of Entomology, London, pp. 140–3 and 200–2.

21 M255.

22 Pozerski de Pomiane, E., 'Souvenirs d'un semi-siècle à l'Institut Pasteur', p. 45; located in the archive of the Pasteur Institute (Fonds E. Pozerski), and M484. Summers mentions the reference in Pozerski's memoirs (see note 1).

23 Bruynoghe, R. and Maisin, J. (1921) 'Essais de thérapeutique au moyen du bactériophage du staphylocoque', Comptes rendus de la Société de Biologie, **85**: 1120–1.

24 Gratia, A. (1922) 'La lyse transmissible du staphylocoque. Sa production; ses applications thérapeuthiques', Comptes rendus de la Société de Biologie, **86**: 276–8.

25 Beckerich, A. and Hauduroy, P. (1922) 'Le bactériophage dans le traitement de la fièvre typhoïde', Comptes rendus de la Société de Biologie, **86**: 168–70.

26 Otto, R. and Munter, H. (1921) 'Zum d'Herelleschen Phänomen', Deutsche Medizinische Wochenschrift, **47**: 1579–82.

27  Zdansky, E. (1925) 'Versuche einer Bakteriophagentherapie bei Coli-Infektionen der abführenden Harnwege', *Wiener Archiv für Innere Medizin*, **11**: 533–48.

28  Raettig, H. (1958/1967) *Bakteriophagie* and *Bakteriophagie 1957–1965*, Gustav Fischer.

29  Rice, T. B. (1930) 'Use of bacteriophage filtrates in treatment of suppurative conditions: report of 300 cases', *American Journal of the Medical Sciences*, **179**: 345–60.

30  Da Costa Cruz, J. (1924) 'Le traitement des dysentéries bacillaires par le bacteriophage', *Comptes rendus de la Société de Biologie*, **91**: 845–6.

31  'Tiny and deadly bacillus has enemies still smaller', *New York Times*, 27 September 1925; Kruif, P. de (1931) 'Miracles of healing', *Ladies Home Journal*, June.

32  Porter, R. (1997) *The Greatest Benefit to Mankind*, Norton & Company, p. 441.

33  Geison, G. L. (1995) *The Private Science of Louis Pasteur*, Princeton University Press, pp. 146–9.

34  Quoted by A. Raiga in *Nouvelles archives hospitalières* (1973) **1**: 11.

35  D'Herelle, F. (1921) *Le bactériophage – son rôle dans l'immunité*, Masson, p. 10.

36  M496–7.

37  The information on the Nobel Prize nominations from the Nobel archives was kindly provided by the Nobel Committee for Physiology or Medicine.

38  The letter is located in the d'Herelle collection in the Pasteur Institute archives. Information on the period in Egypt: M511–33.

39  D'Herelle, F. (1925) 'Essai de traitement de la peste bubonique par le bacteriophage', *Presse Médicale*, **33**: 1393–4. Reprinted with kind permission of Editions Masson.

40  The letter is from a dossier (L/E 1425 file 7616) of the economic and overseas department of the British government. It was found by W. Summers (see note 1). It is currently located in the Oriental and Indian Offices collection of the British Library in London.

41  *Annual Report of the Haffkine Institute (Bombay) (1927)* pp. 35–7, located in the Oriental and Indian Offices collection of the British Library (file V/24/412); Naidu, B. P. B. and Avari C. R. (1932) 'Bacteriophage in the treatment of plague', *Indian Journal of Medical Research*, **19**: 737.

42  M430.

43  Pollitzer, R. (1959) *Cholera*, WHO.

44  See d'Herelle, F. (1926) *Le bactériophage et son comportement*, Masson.

45  A three-page report describes d'Herelle's cholera experiments in India. It is located in dossier L/E 1425 file 7616, pp. 120–43 (see note 40). Additional information: d'Herelle, F. (1928) 'Le cholera asiatique', *Presse Médicale*, **61**: 961–4 and M600–61.

46   'The d'Herelle bacteriophage in the treatment of cholera', *Journal of the American Medical Association*, (1928) **90**: 783–4.

47   *Annual Review of the Public Health Commissioner with the Governor of India (1932)* p. 41, in the Oriental and Indian Office collection of the British Library (file V/24/3661).

48   *Annual Report of the King Edward VII Memorial Pasteur Institute and Medical Research Institute (Shillong) (1928)* p. 3, in the Oriental and Indian Office collection of the British Library (file V24/795/96). Information on the distributed phage vials appeared in *Annual Public Health Report of the Province of Assam (1928–1935)* (file V24/3870/71).

49   Editorial (1930) 'The cholera bacteriophage', *Indian Medical Gazette*, **65**: 91–3.

50   Morison, J. (1932) *Bacteriophage in the treatment and prevention of cholera*, contains the information on the experiments in Assam. The population of Assam and additional details are available in *Annual Public Health Report of the Province of Assam (1928–1935)* (file V24/3870/71).

51   Morison, J. (1935) 'Bacteriophage and cholera', *Transactions of the Royal Society for Tropical Medicine and Hygiene*, **28**: 563–70. Also contains the results in Assam starting in 1932.

52   In fact, WHO-sponsored phage therapy studies with cholera patients were carried out in Pakistan in the late 1960s and early 1970s. They demonstrated some positive effects that were regarded by the researchers as not sufficiently significant. See Monsur, K. A. et al. (1970) 'Effect of massive dosis of bacteriophage on excretion of vibrios, duration of diarrhoea and output of stools in acute cases of cholera', *Bulletin of the World Health Organization*, **42**: 723–32 and Marcuk, L. M. et al. (1971) 'Clinical studies of the use of bacteriophage in the treatment of cholera', *Bulletin of the World Health Organization*, **45**: 77–83.

53   Faruque, S. M. et al. (2005) 'Self-limiting nature of seasonal cholera epidemics: Role of host-mediated amplification of phage', *Proceedings of the National Academy of Sciences of the United States of America*, **102**: 6119–24 and Faruque, S. M. et al. (2005) 'Seasonal epidemics of cholera inversely correlate with the prevalence of environmental cholera phages', *Proceedings of the National Academy of Sciences of the United States of America*, **102**: 1702–7.

54   The German Bacteriophage Society is mentioned in Pockels, W. (1927) 'Die Bakteriophagentherapie in der Kinderheilkunde', *Monatsschrift für Kinderheilkunde* **35**: 229-236; the German Enterofagos advertisement: *Wiener Klinische Wochenschrift* (1943) **56**: 693; the English Enterofagos advertisement appeared in a four-page advertising pamphlet located in the Frederick W. Twort collection at the Wellcome library for the history and understanding of medicine in London (file GC/176, B1).

55   M735–6; Mazure, personal communication; Peitzmann, S. J. (1969) 'Félix
     d'Herelle and bacteriophage therapy', *Transactions and Studies of the
     College of Physicians of Philadelphia*, **37**: 115–23.
56   Straub, M. E. and Applebaum, M. (1932) 'Studies of commercial bacte-
     riophage products', *Journal of the American Medical Association*, **100**:
     110–13; physicians' information booklets published by Parke, Davis
     ('Therapeutic notes', November 1934) and Squibb ('Squibb memo-
     randa', September 1931).
57   Kabeshima, T. (1920) 'Sur un ferment d'immunité bactériolysant, du
     méchanisme d'immunité infectieuse intestinale, de la nature du dit
     'microbe filtrant bactériophage' de d'Herelle', *Comptes rendus de la
     Société de Biologie*, **83**: 219–21, and ibid. 'Sur le ferment d'immunité
     bactériolysant', 471–3.
58   The first phage publication by Bordet; Bordet, J. and Ciuca, M. (1920)
     'Exsudats leucocytaires et autolyses microbiennes transmissibles',
     *Comptes rendus de la Société de Biologie*, **83**: 1293–6; publication in which
     Twort's publication was unearthed: Bordet, J. and Ciuca, M. (1921)
     'Remarques sur l'histoire des recherches, concernant la lyse microbienne
     transmissible', *Comptes rendus de la Société de Biologie*, **84**: 745–7; for
     Twort's publication, see note 10.
59   Nicolle, P. (1949) 'Chroniques – Félix d'Herelle', *Presse Médicale*, **57**: 350.
60   Summarized in d'Herelle, F. et al. (1922) 'Discussion on the bacterio-
     phage (Bacteriolysin). From the ninetieth annual meeting of the British
     Medical Association, Glasgow, July, 1922', *British Medical Journal*, **2**:
     289–97.
61   Gratia, A. (1931) 'Sur l'identité du phénomène de Twort et la bactério-
     phagie (dernière reponse à M. d'Herelle)', *Annales de l'Institut Pasteur*, **47**:
     243–4. The letter is cited in Twort, A. (1993) *In Focus, Out of Step – a Biog-
     raphy of Frederick William Twort F. R. S. 1877–1950*, Allan Sutton, p. 200.
     The letter is located in the Twort collection (see note 54).
62   D'Herelle's challenge to a duel: d'Herelle, F. (1931) 'Le phénomène de
     Twort et la bacteriophagie', *Annales de l'Institut Pasteur*, **47**: 241–2. His
     court order appeared in same year and volume on pages 470–1. The
     verdict: Flu, P. C. and Renaux, E. (1932) 'Le phénomène de Twort et la
     bactériophagie', *Annales de l'Institut Pasteur*, **48**: 15–18.
63   'The use of bacteriophage', *Science*, New Series **70**(1817): x.
64   See note 31.
65   'Our contemporary aces', *Science*, New Series 71(1840): 361; Hauduroy,
     P. (1931) 'La thérapeutique par le bactériophage: ses avantages, ses
     dangers', *Press Médicale*, **10**: 168–71.
66   MacNeal, W. J. (1934) 'Using the enemy of bacteria to combat disease',
     *Literary Digest*, **117**: 17.

67   See note 56 (Straub) and Rakieten, M. L. (1932) 'Studies with *staphylo-coccus* bacteriophage I. The preparation of polyvalent *staphylococcus* bacteriophage', *Yale Journal of Biology and Medicine*, **4**: 807–19.

68   Polemics against phage therapy: editorials in *Journal of the American Medical Association*, (1931) **96**: 693, (1932) **98**: 1190, (1933) **100**: 1431–2 and 1603–4; Eaton, M. D. and Bayne-Jones, S. (1934) 'Bacterio-phage therapy: review of the principles and results of the use of bacteriophage in the treatment of infections (I-III)', *Journal of the American Medical Association*, **103**: 1769–76, 1847–53, 1934–9.

69   Fisk, R. T. (1938) 'Protective action of typhoid phage on experimental typhoid infection in mice', *Proceedings of the Society for Experimental Biology and Medicine*, **38**: 659–60.

70   See note 32, p. 529.

71   *Annual Report of the King Edward VII Memorial Pasteur Institute and Medical Research Institute (Shillong) (1936)* p. 13, in the Oriental and Indian Office collection of the British Library (file V24/795/96).

72   Phage therapy was also used in other areas of India: *Annual Public Health Report of the Province of Assam (1938)* p. 3 (file V24/3870/72–74); Pasricha, C. L. et al. (1936) 'Bacteriophage in the treatment of cholera'; *Indian Medical Gazette*, **71**: 61–8; 'More about phage', *Lancet*, 15 Nov 1941, pp. 607–8.

73   Krueger, A. P. and Scribner, E. J. (1941) 'The bacteriophage: Its nature and its therapeutic use (I/II)' *Journal of the American Medical Association*, **116**: 2160–7, 2269–77.

# Chapter 4

1   On the insufficient treatment options, see Chapter 3, note 6; on tetanus antitoxins, see Chapter 3, note 32, p. 443.

2   On the treatment of dysentery, see Chapter 3, note 6; on prontosil, see Chapter 3, note 32, pp. 453–4.

3   Gantenberg, R. (1939) 'Ruhr aus dem Feldzug in Polen', *Deutsche Medizinische Wochenschrift*, **65**: 1769–93, 1820, 1825.

4   The information about polyfagin is from the archives of the former Höchst AG. The archives are managed by HistoCom, who kindly made them available.

5   Guleke, N. (1941) 'Soll die frische Hirnschußwunde genäht werden?', *Der Deutsche Militärarzt*, **6**: 157.

6   See Chapter 3, note 6.

7   Klose, F. and Schröer, W. (1941) 'Ein Beitrag zur Ruhrschutzbehandlung mit polyvalenten Ruhrbakteriophagen', *Der Deutsche Militärarzt*, **6**: 265–7. The location of the POW camp is not mentioned in the report.

8   Ibid.

9   Mazure, personal communication.

10  Boyd, J. S. K. and Portnoy, B. (1944) 'Bacteriophage therapy in bacillary dysentery', *Transactions of the Royal Society of Tropical Medicine and Hygiene*, **37**: 243–62.

11  Jadin, J. and Resseler, J. (1957) 'La dysenterie bacillaire au Ruanda-Urundi et au Kivu', *Annales de la Société Belge de Médecine Tropicale*, **37**: 347–69.

12  See Chapter 3, note 6.

13  Letter from M. Rakieten to F. d'Herelle dated 30 April 1942, located in the d'Herelle collection in the Pasteur Institute archives (see Chapter 3, note 1).

14  Dubos, R. J. et al. (1943) 'The multiplication of bacteriophage in vivo and its protective effects against an experimental infection with *Shigella dysenteriae*', *Journal of Experimental Medicine*, **78**: 161–8.

15  Morton, H. E. and Perez-Otero, J. E. (1945) 'The increase of bacteriophages in vivo during experimental infections with *Shigella paradysenteriae* in mice', *Journal of Bacteriology*, **49**: 237–44; Morton, H. E. and Engley, F. B. (1945) 'The protective action of dysentery bacteriophages in experimental infections in mice', *Journal of Bacteriology*, **49**: 245–55.

16  Ibid, p. 245.

17  Schade, A. L. and Caroline, L. (1943) 'The preparation of a polyvalent dysentery bacteriophage in a dry and stable form', *Journal of Bacteriology*, **46**: 463–73, and ibid. (1944) **48**: 179–90, 243–51.

18  Morton, H. E. and Engley, F. B. (1945) 'Dysentery bacteriophage', *Journal of the American Medical Association*, **127**: 584–91.

19  Ward, W. E. (1943) 'Protective action of Vi bacteriophage in *Eberthella typhi* infections in mice', *Journal of Infectious Diseases*, **72**: 172–6; on the first Vi phages: Craigie, J. and Brandon, K. F. (1936) 'The laboratory identification of the V form of *B. typhosus*', *Canadian Journal of Public Health*, **27**: 165.

20  Knouf, E. G. et al. (1946) 'Treatment of typhoid fever with type-specific bacteriophage', *Journal of the American Medical Association*, **132**: 134–8; Desranleau, J.-M. (1949) 'Progress in the treatment of typhoid fever with Vi bacteriophages', *Canadian Journal of Public Health*, **40**: 473.

21  Corbel, M. J. and Morris, J. A. (1980) 'Investigation of the effect of Brucella-phage on the course of experimental infection with *Brucella abortus*', *British Veterinary Journal*, **136**: 278–89.

22  Broxmeyer, L. et al. (2002) 'Killing of *Mycobacterium avium* and *Mycobacterium tuberculosis* by a mycobacteriophage delivered by a nonvirulent mycobacterium: a model for phage therapy of intracellular bacterial pathogens', *Journal of Infectious Diseases*, **186**: 1155–60. A second group led by Pablo Bifani, a researcher at the Pasteur Institute in Brussels who

worked with the US-British biotech company PhageGen, uses a different list. Bifani and the PhageGen scientists coupled small proteins with the phages. These proteins are supposed to get the macrophages to take up the phages. This type of penetration protein is produced by HIV viruses, for example, that use them for the same purpose. Bifani, personal communication.

23 Raiga, A. (1978) *Nouvelles Archives Hospitalière*, 6: 31.

24 Bertschinger, J. P. (1957) 'Le bactériophage', *Schweizerische Apotheken-zeitung*, **95**: 479–87; 'Das neue Arzneipräparat' *Medizinische Monatsschrift*, **13**: 124 (1959); and information provided by Guy and Michel-Pierre Glauser, the two sons of deceased Saphal owner Hermann Glauser, as well as Jean-Pierre Feihl, who was a pupil of Hauduroy and used phages in the 1950s.

25 Mazure, personal communication.

26 See Chapter 3, note 1.

27 See Chapter 3, note 37.

28 Mazure, personal communication and Raiga, A. (1949) 'Félix d'Herelle 1873–1949', *Vie médicale*, **5**: 37–8.

29 Mullins, N. C. (1974) 'Die Entwicklung eines wissenschaftlichen Spezial-gebiets: die Phagen-Gruppe und die Ursprünge der Molekularbiologie', in Weingart, P. (ed.) *Wissenschaftssoziologie 2*, Fischer Athenäum, pp. 184–222.

30 Arber, W. (1978) 'Promotion and limitation of genetic exchange', Nobel lecture.

31 See Chapter 3, note 28.

32 Wommack, K. E. and Colwell, R. R. (2000) 'Virioplankton: viruses in aquatic ecosystems', *Microbiology and Molecular Biology Reviews*, **64**: 69–114; Copley, J. (2002) 'All at sea', *Nature*, **415**: 572–4.

33 Kenney, J. E. and Bitton, G. (1987) 'Bacteriophages in food', in Goyal, S. M. et al. (eds) *Phage Ecology*, John Wiley & Sons, pp. 289–316.

34 Ackermann, H.-W. (2001) 'Frequency of morphological phage descrip-tions in the year 2000', *Archives of Virology*, **146**: 843–7.

35 See note 32.

36 'Scientists destroy butyl alcohol foes', *New York Times*, 6 May 1944.

# Chapter 5

1 Georgadze, I. (1974) 'Fifty years of Tbilisi Research Institute for vaccine and serum', in *Theoretical and Practical Aspects of Bacteriophages*, Tbilisi, pp. 5–58 (in Georgian, English translation by M. D. Serebryakova).

2 In addition to Nina Kilasonidze, Eliava's employees who passed on their know-how included Elena Makashvili and Irakli Georgadze, who later

became the director of the institute. Eliava's stepdaughter Hanna Maliev died in 1997.

3  Tarkhan-Mouravi, G., '70 years of Soviet Georgia – From independence to independence: 1921–1991', unpublished manuscript.

4  Much of what is known about Eliava's life is based on stories passed down from his stepdaughter Hanna Maliev. They were retold by Maliev's daughter Nathalia Devdariani and Nino Chanishvili of the Eliava Institute, who knew Maliev and interviewed her several times. Biographical information in particular was taken from the source in note 1. According to Devdariani, Hanna Maliev was the daughter of Eliava's wife Amelia Vol-Levitskaya and her first husband, Mikolai Lewicki.

5  Eliava's studies at the universities of Odessa (1909–1910) and Geneva (1912–1914) are confirmed by the University of Geneva's enrolment records (archivist Dominique Anne Torrione-Vouilloz, personal communication on 5 December 2002).

6  See note 1.

7  See note 1 and Danelia, F. (1992) 'Georgiy Eliava', *Georgian Medical Magazine* No. 2. This article describes two written recommendations, which confirm that Eliava was sent to Paris by the Georgian government for the purposes of education and research. The recommendation dated 31 October 1919 was from the foreign ministry and was addressed to the Georgian embassy in Paris, and the other one was from a deputy minister without a direct addressee (in Georgian, English translation by M. D. Serebryakova). Facsimiles of the documents are in the possession of N. Devdariani.

8  The Mtkvari River is also known by its Russian name 'Kura'.

9  Kryvinsky, A. S. (1962) *Viruses versus Microbes*, State Medical Publisher Moscow, cited in Sherbina-Esvandjia, L., 'Georgia needs me' from the newspaper *Zaria Vostokova*, 23 June 1988 (Kryvinsky's article is in Russian, Sherbina-Esvandjia's is in Georgian. English translation by M. D. Serebryakova); Ermolyeva, Z. V. and Yakobson, L. M. (1945) 'The diagnosis of cholera and the effect of bacteriophage prophylaxis during cholera outbreaks', in Babsky, E. B. et al. (eds) *Microbiology and Epidemiology – Achievements of Soviet Medicine in the Patriotic War*, Medical Publications (English edition), p. 53, the Russian original was published in 1943.

10  See Chapter 3, note 22.

11  Letter dated 27 February (year not given) from Eliava to Edouard Dujardin-Beaumetz, who worked at the Pasteur Institute, located in the Dujardin-Beaumetz collection in the archives of the Pasteur Institute.

12  Elie Wollman, personal communication and recording of a round table discussion that Wollman participated in. The cassette is located in the d'Herelle collection of the Pasteur Institute archives (see Chapter 3, note 1).

13 Sherbina-Esvandjia, L. (1988) 'Georgia needs me', in *Zaria Vostokova*, 23 June (in Georgian, English translation by M. D. Serebryakova).
14 Ibid.
15 See note 1.
16 See note 3.
17 See note 1.
18 See Chapter 3, note 22.
19 N. Chanishvili, personal communication.
20 Mazure, personal communication.
21 Summers, W. C. (1999) *Félix d'Herelle and the Origins of Molecular Biology*. Yale University Press, p. 159.
22 See note 11. Regarding the date of the letter: it is most probably from the period between 1928 and 1933, since d'Herelle's Laboratoire du Bactério-phage, whose advertising Eliava was writing about, was founded in 1928. Emile Roux, the director of the Pasteur Institute, who is also mentioned in the letter, died in 1933. My thanks go to N. Chanishvili for pointing out this letter to me.
23 For example, Osborne, L. (2000) 'A Stalinist antibiotic alternative', *New York Times Magazine*, 6 March; Carlton, R. M. (1999) 'Phage therapy: past history and future prospects', *Archivum Immunologiae et Therapiae Experimentalis* (Warsz) **47**: 267–74.
24 D'Herelle, F. (1935) *Bacteriophag i fenomen vyzdorovleniya*, Tbilisi National University (translated from the German translation by Y. Scherrer).
25 M11, 161 and 301.
26 See note 24.
27 M770.
28 M191–4.
29 M685.
30 M168.
31 Quoted in Gunther, J. (1957) *Inside Russia Today*, Harper, p. 253.
32 D'Herelle, F., 'La valeur de l'expérience', manuscript in the d'Herelle collection of the Pasteur Institute archives (see Chapter 3, note 1) p. 3.
33 See note 31, p. 277.
34 Cited in source in note 1.
35 Information on his staff provided by Kilasonidze, personal communication; on the honorary doctorate, see note 1.
36 Loose notes in the d'Herelle's collection of the Pasteur Institute archives (see Chapter 3, note 1).
37 See note 24, p. 8.
38 Kilasonidze's statements correspond to the descriptions in the source mentioned in note 1.
39 Vronskaya, J. and Chuguev, V. (2000) *The Biographical Dictionary of the Former Soviet Union 1917–1988*, Bowker/Saur.

40  Ibid.
41  Information on Mdivani in the source in note 3; Kilasonidze stated that Eliava had access to him.
42  See note 13.
43  See note 1.
44  See note 36.
45  Mazure, personal communication.
46  See note 31, p. 142.
47  Bullock, A. (1991) *Hitler and Stalin – Parallel Lives*. HarperCollins, p. 517.
48  Cited in Bullock (see note 47) p. 517.
49  See note 1.
50  Mazure, personal communication.
51  Gazaryan, S. (1989) 'It must not happen again', Zvezda 1: 3, cited in Shrayer, D. P. (1996) 'Félix d'Herelle in Russia', *Bulletin de l'Institut Pasteur*, **94**: 91–6.
52  See note 1.
53  Mazure, personal communication.
54  Pyatokov, cited in Bullock (see note 47) p. 531; Beria, cited in Tarkhan-Mouravi (see note 3).
55  See note 39.
56  Cited in Bullock (see note 47) p. 560.
57  See note 9.
58  See notes 7 and 9.
59  See for example, 'Plot with Reich and Japan confessed at Soviet trial', *New York Times*, 24 January 1937.
60  Ordzhonikidze: see note 47; Mdivani: see note 3.
61  N. Devdariani, personal communication (see note 4).
62  Copy of a newspaper report in the Edouard Pozerski collection in the archives of the Pasteur Institute.
63  Cited with the kind permission of G. Tarkhan-Mouravi (see note 3).
64  Bullock, p. 552 (see note 47) and Gunther, p. 143 (see note 33).
65  See note 1.
66  Ibid.
67  Ibid.
68  Krestovnikova, V. A. (1947) 'Phage therapy and phage prophylaxis and their justification in the studies of Soviet researchers', *Zhurnal mikrobiologii, epidemiologii i immunobiologii*, N3: 56–65; (in Russian, German translation by Y. Scherrer); see also Rogozin, I. I. (1945) 'The institutes of epidemiology and microbiology during the patriotic war', in Babsky et al. (see note 9).
69  Cf. ibid. and Merril, C. et al. (1996) 'Long-circulating bacteriophages as antibacterial agents', *Proceedings of the National Academy of Sciences of the United States of America*, **93**: 3188–92.

70   Ibid, see also note 1.
71   Ibid.
72   Belikov, P. E. (1947) 'The fight against intestinal infections', *American Review of Soviet Medicine*, **4**: 238—42.
73   Zhuravlev, P. M. and Pokrovskaya, M. P. (1945) 'The phagoprophylaxis and phagotherapy of gas gangrene', in Babsky et al. (see note 9).
74   Lebed, A. I. et al. (eds) (1966) *Who's Who in the USSR 1965–1966*, 2nd edn, Scarecrow Press.
75   See note 68.
76   Zhuravlev, see note 73.
77   Statistics on victims in Georgia taken from Tarkhan-Mouravi (see note 3); overall statistics: Living Museum Online of the German Historical Museum: www.dhm.de/lemo/home.html (German).
78   Waksman, S. A. (1947) 'Microbiology in the USSR in 1946', *Scientific Monthly*, **64**: 289–96.
79   Today, Sverdlovsk is called Yekaterinburg and Stalinabad is called Dushanbe.
80   Sergienko, F. E. (1945) 'Dry bacteriophages, their preparation and use', in Babsky et al. (see note 9).
81   Babsky et al., p. 156 (see note 9).
82   Mudd, S. (1947) 'Recent observations on programs for medicine and national health in the USSR', *Science*, **105**: 269–73, 306–9.
83   Gunther, pp. ixi, 271, 279 (see note 31); Belikov, p. 241 (see note 72); D. P. Shrayer, personal communication.
84   T. Chanishvili and Victor Krylov, a Russian phage researcher, personal communication.
85   Today called Nizhny Novgorod.
86   See note 1.
87   See note 74.
88   See note 68.
89   Ramesh, V. et al. (1999) 'Prevention of *Clostridium difficile*-induced iliocecitis with bacteriophage', *Anaerobe*, **5**: 69–78.
90   Bartlett, J. G. (2002) 'Antibiotic-associated diarrhoea', *New England Journal of Medicine*, **346**: 334–9.
91   *The Merck Manual of Diagnosis and Therapy*, 17th edn, John Wiley & Sons, 1999; Yao, J. D. C. and Moellering, R. C. (1995) 'Antimicrobial agents', in Murray, P. R. et al. (eds) *Manual of Clinical Microbiology*, 7th edn, American Society for Microbiology, pp. 1281–307.
92   Cited in Alisky, J. et al. (1998) 'Bacteriophages show promise as antimicrobial agents', *Journal of Infection*, **36**: 5–15.
93   Shrayer, personal communication and Shrayer, D. P. (1989) *Staphylococcal Disease in the Soviet Union. Epidemiology and Response to a National Epidemic*, Delphic Associates.

94  See note 1.
95  Liu, M. et al. (2002) 'Reverse transcriptase-medicated tropism switching in *Bordetella* bacteriophage', *Science*, **295**: 2091–4.
96  Vieu, personal communication; Vieu, J.-F. (1975) 'Les bactériophages', in Fabre, J. (ed.) *Traité de Thérapeutique, Vol. Serums et Vaccins*, Flammarion, pp. 337–40; Vieu, J.-F. et al. (1979) 'Données actuelles sur les applications thérapeutiques des bactériophages', *Bulletin de l'Academie Nationale de Medecine*, **163**: 61–5.
97  T. Chanishvili, personal communication.
98  See note 1.

# Chapter 6

1  Meipariani, personal communication.
2  T. Chanishvili, personal communication.
3  Radetsky, P. (1996) 'Return of the good virus', *Discover*, **17**: 50–8.
4  BBC documentary film 'The virus that cures'; Kutter, personal communication.
5  N. Chanishvili, personal communication.
6  BBC documentary film 'The virus that cures'; Ackermann, personal communication.
7  Katsarava, personal communication; Stone, R. (2002) 'Stalin's forgotten cure', *Science*, **298**: 728–31; Jikia, D. (2005) 'The use of a novel biodegradable preparation capable of the sustained release of bacteriophages and ciprofloxacin, in the complex treatment of multidrug-resistant *Staphylococcus aureus*-infected local radiation injuries caused by exposure to Sr90', *Clinical and Experimental Dermatology*, **30**: 236.
8  Markoishvili, K. et al. (2002) 'A novel sustained-released matrix based on biodegradable polyester amides and impregnated with bacteriophages and an antibiotic shows promise in management of infected venous stasis ulcers and other poorly healing wounds', *International Journal of Dermatology*, **41**: 453–8.
9  Alisky, J. et al. (1998) 'Bacteriophages show promise as antimicrobial agents', *Journal of Infection*, **36**: 5–15; Sulakvelidze, A. et al. (2001) 'Bacteriophage therapy', *Antimicrobial Agents and Chemotherapy*, **45**: 649–59; Chanishvili et al. (2001) 'Phages and their application against drug-resistant bacteria', *Journal of Chemical Technology and Biotechnology*, **76**: 1–11.
10  Kutter and Gvasalia, personal communication.
11  Waldor, M. K. and Mekalanos, J. J. (1996) 'Lysogenic conversion by a filamentous phage encoding cholera toxin', *Science*, **272**: 1910–14.

12   Köhler, B. et al. (2000) 'Antibacterials that are used as growth promoters in animal husbandry can affect the release of Shiga-toxin-2-converting bacteriophages and Shiga toxin 2 from *Escherichia coli* strains', *Microbiology*, **146**: 1085–90.

13   Krause, R. M. (1992) 'The origins of plagues: old and new', *Science*, **257**: 1073–8.

14   Kutter, personal communication.

## Chapter 7

1   Danelia, N. et al., 'Bakteriophagen zur Therapie von MRSA-Infektionen', unpublished study; Danelia, N. et al. (1998) 'Bakteriophagen zur Therapie von multi-resistenten Bakterien', *Zeitschrift für Wundbehandlung*, **3**: 16–17.

2   Bitter-Suermann, personal communication.

3   Weber-Dabrowska, B. et al. (2000) 'Bacteriophage therapy of bacterial infections: an update of our institute's experience', *Archivum immunologiae et therapiae experimentalis* (Warsz) **48**: 547–51.

4   Results communicated to the author earlier were recanted by the same source, who claimed that he had lied in order not to jeopardize the search for investors.

5   Biswas, B. et al. (2002) 'Bacteriophage therapy rescues mice bacteremic from a clinical isolate of vancomycin-resistant *Enterococcus faecium*', *Infection and Immunity*, **70**: 204–10.

6   Hancock, D. (2000) 'More *E. coli*?', press release, Washington State University, 10 March; Becker, E. (2002) '19 million pounds of meat recalled after 19 fall ill', *New York Times*, 20 July.

7   Health Protection Agency (UK) accessible at www.hpa.org.uk/infections/topics_az/list.htm.

8   Ibid.

9   Tiwana, J. (2001) '*Salmonella* falls in chicken; *Campylobacter* rises', *Food Chemical News*, **43**: 28.

10   'Superbugs found in chicken survey', BBC news online, 16 August 2005.

11   See for instance Huff, W. E. et al. (2005) 'Alternatives to antibiotics: utilization of bacteriophage to treat colibacillosis and prevent foodborne pathogens', *Poultry Science*, **84**: 655–9.

12   Barrow, personal communication. As Smith's successor, he carried out several phage therapy trials himself: Barrow, P. A. et al. (1998) 'Use of lytic bacteriophage for control of experimental *Escherichia coli* septicaemia and meningitis in chickens and calves', *Clinical and Diagnostic Laboratory Immunology*, **5**: 294–8; Berchieri, A. et al. (1991) 'The activity in the

chicken alimentary tract of bacteriophages lytic for *Salmonella typhimurium*', *Research in Microbiology*, **142**: 541–9.

13 Smith, H. W. and Huggins, M. B. (1982) 'Successful treatment of experimental *Escherichia coli* infections in mice using phage: its general superiority over antibiotics', *Journal of General Microbiology*, **128**: 307–18.

14 Merril, C. R. et al. (2003) 'The prospect for bacteriophage therapy in Western medicine', *Nature Reviews Drug Discovery*, **2**: 489–97.

15 Smith, H. W. and Huggins, M. B. (1983) 'Effectiveness of phages in treating experimental *E. coli* diarrhoea in calves, piglets and lambs', *Journal of General Microbiology*, **129**: 2659–75; Barrow, P. A. (2001) 'The use of bacteriophages for treatment and prevention of bacterial disease in animal and animal models of human infection', *Journal of Chemical Technology and Biotechnology*, **76**: 677–82.

16 Smith, H. W. and Huggins, M. B. (1987) 'The control of experimental *E. coli* diarrhoea in calves by means of bacteriophage', *Journal of General Microbiology*, **133**: 111–26.

17 Kutter, personal communication.

18 See note 16.

19 Ramachandran, personal communication.

20 Lecture by A. Sulakvelidze, Intralytix, at the congress of the International Union of Microbiological Societies held on 1 August 2002 in Paris; the lecture was also the source for some of the data presented later in the chapter.

21 Sulakvelidze, personal communication; 'Quantitative assessment of relative risk to public health from foodborne *Listeria monocytogenes* among selected categories of ready-to-eat foods', FDA, USDA and CDC report from September 2003, available at: www.cfsan.fda.gov/~dms/lmr2-toc.html; Leverentz, B. et al. (2003) 'Biocontrol of *Listeria monocytogenes* on fresh-cut produce by treatment with lytic bacteriophages and a bacteriocin', *Applied and Environmental Microbiology*, **69**: 4519–26.

22 Flaherty, J. E. et al. (2000) 'Control of bacterial spot on tomato in the greenhouse and field with h-mutant bacteriophage', *HortScience*, **35**: 882–4.

23 C. Snowdon, Omnilytics, personal communication.

24 Levin, B. R. and Bull J. J. (2004) 'Population and evolutionary dynamics of phage therapy', *Nature Reviews Microbiology*, **2**: 166–73.

25 Chibani-Chennoufi, S. et al. (2004) 'In vitro and in vivo bacteriolytic activities of *Escherichia coli* phages: implications for phage therapy', *Antimicrobial Agents and Chemotherapy*, **48**: 2558–69; Chibani-Chennoufi, S. et al. (2004) 'Isolation of *Escherichia coli* bacteriophages from the stool of paediatric diarrhoea patients in Bangladesh', *Journal of Bacteriology*, **186**: 8287–94.

26  Bruttin, A. and Brüssow, H. (2005) 'Human volunteers receiving *Escherichia coli* phage T4 orally; a safety test of phage therapy', *Antimicrobial Agents and Chemotherapy*, **49**: 2874–8.

27  Merril, C. R. et al. (1971) 'Bacterial virus gene expression in human cells', *Nature*, **233**: 398–400.

28  Geier, M. R. et al. (1973) 'The fate of bacteriophages as antibacterial agents', *Nature*, **246**: 221–3.

29  Merril, C. R. et al. (1996) 'Long-circulating bacteriophages as antibacterial agents', *Proceedings of the National Academy of Sciences of the United States of America*, **93**: 3188–92.

30  School, D. et al. (2001) 'Bacteriophage K1-5 encodes two different tail fibre proteins, allowing it to infect and replicate on both K1 and K strains of *Escherichia coli*', *Journal of Virology*, **75**: 2509–15.

31  Two research groups have recently shown that genetically modified phages that can no longer multiply in the bacterium do not need additional toxic genes. Infesting the bacteria is sufficient to cure mice that have been infected by them. The researchers hope that with this approach the bacteria in the site of infection are dissolved more slowly than is the case with phages that are able to multiply. In so doing, the bacterial debris that drives the immune system to dangerous overreactions is released more slowly. In actual fact, the researchers observed less strong inflammatory responses in mice treated in this way: Hagens, S. et al. (2004) 'Therapy of experimental *Pseudomonas* infections with a nonreplicating genetically modified phage', *Antimicrobial Agents and Chemotherapy*, **48**: 3817–22, and Matsuda, T. et al. (2005) 'Lysis-deficient bacteriophage therapy decreases endotoxin and inflammatory mediator release and improves survival in a murine peritonitis model', *Surgery*, **137**: 639–46. Norris's experiments: Norris, J. S. et al. (2000) 'Procaryotic gene therapy to combat multidrug resistant bacterial infection', *Gene Therapy*, **7**: 723–5; Westwater, C. et al. (2003) 'Use of genetically engineered phage to deliver antimicrobial agents to bacteria: an alternative therapy for treatment of bacterial infections', *Antimicrobial Agents and Chemotherapy*, **47**: 1301–7.

32  Loeffler, J. M. et al. (2001) 'Rapid killing of *Streptococcus pneumoniae* with a bacteriophage cell wall hydrolase', *Science*, **294**: 2170–2; Nelson, D. et al. (2001) 'Prevention and elimination of upper respiratory colonization of mice by group A streptococci by using a bacteriophage lytic enzyme', *Proceedings of the National Academy of Sciences of the United States of America*, **98**: 4107–12. Martin J. Loessner of the Institute of Food Science and Nutrition of the Swiss Federal Institute of Technology in Zurich is pursuing a comparable project. He is investigating the extent to which the lytic enzymes of *Listeria* phages can be used to clean *Listeria* off

cheese rinds: Gaeng, S. et al. (2000) 'Gene cloning and expression and secretion of *Listeria monocytogenes* bacteriophage-lytic enzymes in *Lactococcus lactis*', *Applied Environmental Microbiology*, **66**: 2951–8.

33  Nguyen, L. et al. (1997) 'Molecular epidemiology of *Streptococcus pyogenes* in an area where acute pharyngotonsillitis is endemic', *Journal of Clinical Microbiology*, **35**: 2111–14.

34  Schuch, R. et al. (2002) 'A bacteriolytic agent that detects and kills *Bacillus anthracis*', *Nature*, **418**: 884–9.

35  Yoong P. et al. (2004) 'Identification of a broadly active phage lytic enzyme with lethal activity against antibiotic-resistant *Enterococcus faecalis* and *Enterococcus faecium*', *Journal of Bacteriology*, **186**: 4808–12.

# Chapter 8

1  D'Herelle, F. (1948) 'Le bactériophage', *Atomes*, **3**: 399–403; English translation in *Science News*, London (1949) 14: 44–59.

2  Merril, C. R. et al. (2003) 'The prospect for bacteriophage therapy in Western medicine', *Nature Reviews Drug Discovery*, **2**: 489–97.

3  Oliveira, D. C. et al. (2002) 'Secrets of success of a human pathogen: molecular evolution of pandemic clones of meticillin-resistant *Staphylococcus aureus*', *The Lancet Infectious Diseases*, **2**: 180–9.

4  British doctor James Soothill, inspired by Williams Smith, has carried out interesting preliminary trials for using phages in the treatment of infected burn wounds. In the early 1990s he demonstrated that phages can sanitize wounds that have been infected by pseudomonades, making skin grafts possible: Soothill, J. S. (1994) 'Bacteriophage prevents destruction of skin grafts by *Pseudomonas aeruginosa*', *Burns*, **20**: 209–11. Recently Soothill's group managed to use phages prophylactically for wound infections with staph in animal experiments: Wills, Q. F. et al. (2005) 'Experimental bacteriophage protection against *Staphylococcus aureus* abscesses in a rabbit model', *Antimicrobial Agents and Chemotherapy*, **49**: 1220–1.

5  Levin, B. R. and Bull, J. J. (2004) 'Population and evolutionary dynamics of phage therapy', *Nature Reviews Microbiology*, **2**: 166–73.

6  Ochs, H. D. et al. (1971) 'Immunologic responses to bacteriophage ΦX174 in immunodeficiency diseases', *The Journal of Clinical Investigation*, **50**: 2559–68.

7  Fred Bledsoe died of complications from diabetes seven months after his treatment in Tbilisi. Kutter, personal communication; Nutt, A. E. (2003) 'Germs that fight germs', *Newark Star-Ledger*, 9 December; 'Therapy uses viruses as natural antibiotics', *Seattle Times*, 21 January 2003.

# Appendix 1

1 The list is restricted to the pathogenic bacteria mentioned in the book.

2 The enumeration of the diseases caused by the bacteria often only includes a selection.

3 spp. means that several species are referred to at the same time, for example *Salmonella* spp. for *S. typhi* and *S. paratyphi* and so on.

4 Wagenaar, J. A. et al. (2001) 'Phage therapy of *Campylobacter jejuni* colonization in broilers', *International Journal of Medicine and Microbiology*, **291**: 92–3.

5 Monsur, K. A. et al. (1970) 'Effect of massive doses of bacteriophage on excretion of vibrio, duration of diarrhoea and output of stools in acute cases of cholera', *Bulletin of the World Health Organization*, **42**: 723–32; Marcuk, L. M. et al. (1971) 'Clinical studies of the use of bacteriophage in the treatment of cholera', *Bulletin of the World Health Organization*, **45**: 77–83.

# figure sources

The author and publisher wish to acknowledge the sources and to thank the following for use of copyright material in figures within this book:

Figure 1.1: Professor Elizabeth Kutter, Evergreen College, Olympia, WA, USA; Figure 2.1: Dr Janice Carr, CDC, Atlanta, USA; Figure 2.2: Dr Janice Carr, CDC, Atlanta, USA; Figure 2.3: Dr Markus Dürrenberger, ZMB, Biozentrum, University of Basel, Basel, Switzerland; Figure 3.1: Institut Pasteur, Paris, France; Figure 3.2: Dr Markus Dürrenberger, ZMB, Biozentrum, University of Basel, Basel, Switzerland; Figure 3.3: Dr Markus Dürrenberger, ZMB, Biozentrum, University of Basel, Basel, Switzerland; Figure 3.4: Thomas Fritschi and Rich Weber, Zurich, Switzerland; Figure 3.5: Thomas Fritschi and Rich Weber, Zurich, Switzerland; Figure 3.6: Professor M. V. Parthasarathy, Cornell Integrated Microscopy Center, Cornell University, Ithaca, NY, USA; Figure 3.7: Dr Markus Dürrenberger, ZMB, Biozentrum, University of Basel, Basel, Switzerland; Figure 3.8: Professor Hans-Wolfgang Ackermann, Laval University, Quebec City, QC, Canada; Figure 3.9: Thomas Häusler, Basel, Switzerland; Figure 3.10: Institut Pasteur, Paris, France; Figure 3.11: Wiener Klinische Wochenschrift, Springer, Vienna, Austria; Figure 5.1: Provided by Nino Chanishvili, Tbilisi, Georgia; Figure 5.2: Provided by Amiran Meipariani, Tbilisi, Georgia; Figure 5.3: Provided by Nina Kilasonidze, Tbilisi, Georgia; Figure 5.4: Provided by Amiran Meipariani, Tbilisi, Georgia; Figure 5.5: Provided by Amiran Meipariani, Tbilisi, Georgia; Figure 5.6: Provided by Amiran Meipariani, Tbilisi, Georgia; Figure 5.7: Provided by Amiran Meipariani, Tbilisi, Georgia; Figure 5.8: Provided by Amiran Meipariani, Tbilisi, Georgia; Figure 6.1: Thomas Häusler, Basel, Switzerland; Figure 6.2: Provided by Ramaz Katsarava, Tbilisi, Georgia; Figure 6.3: Thomas Häusler, Basel, Switzerland; Figure 6.4: Thomas Häusler, Basel, Switzerland.

# index

284

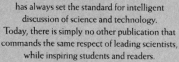

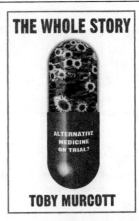
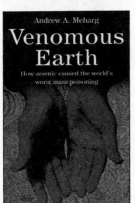

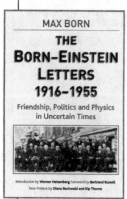